UNDERSEA COLONIES

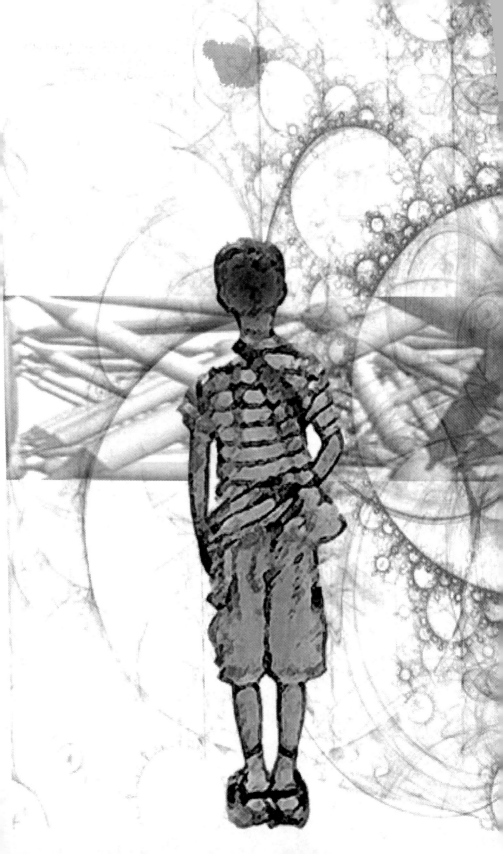

UNDERSEA COLONIES

THE FUTURE OF PERMANENT UNDERSEA SETTLEMENTS

DENNIS CHAMBERLAND

QUANTUM EDITIONS
ORLANDO - CHATTANOOGA

Undersea Colonies

Library of Congress Cataloging-in-Publication Data

Chamberland, Dennis, 1951-
Undersea colonies / Dennis Chamberland.
p. cm.
ISBN 978-1-889422-15-2 (pbk. : alk. paper)
1. Undersea colonies. I. Title.
GC66.5.C43 2007
910.9162--dc22

Printed in the United States of America

Cover Art: Images licensed from Istock Photo, Inc. for this work.

LNW Modeling including the Leviathan and DST II
by Brett W. English.

Maison Sous-Marine – Observatorie
And Village Sous Marin
by Jacques Rougerie

The Hydrofoil and Personal Submarines
by Guillermo Sureda-Burgos

All other images are either in the public domain, used by
permission or are source marked.

Other Books by Dennis Chamberland

Aaron Seven® Adventures

Quantum Storms

Abyss of Space

Fiction

Abyss of Elysium – Mars Wars

Alyete – Dogs of Eros Damned

Exploration Series

The Proxima Manual of Space Exploration with Claudia Chamberland

Undersea Colonies

Pulling the Plug – Real Personal Energy Independence

Children of God Series

Consuming Fire

Proverbs for My Children

The Way Back Home

To the ones who began this great quest:

George Bond and Jacques-Yves Cousteau

To the ones that have and still hold open the doors to the
vast undersea regions for us all:

Ian Koblick and James Miller

To the men who have spent more combined
time there than any others in history:

Christopher Olstad and Richard Presley

To my crewmate, the man who has dared to dive deeper in
more distant and dark places in the flesh than any other:

Terrance Tysall

To the ones who gave me the first opportunity to live there:

Dr. William Knott III and Mark E. Ward

To Aquatica's original daring barnstormers:

Morgan Wells and Bob Barth

And to Aquatica's First Engineer,

Joseph M. Bishop

These, my mentors, heroes and friends,

remain the valiant defenders of the still empty empire.

CORRECTING THE COURSE...

The First Law of Human Exploration
And The First Principle of Undersea Colonization

"In every exploration system, we must require the systems we build to adapt to the human standard rather than expect the human to adapt to the machine or the environment – and in every design activity we will protect the human as a primary objective."

Dennis Chamberland

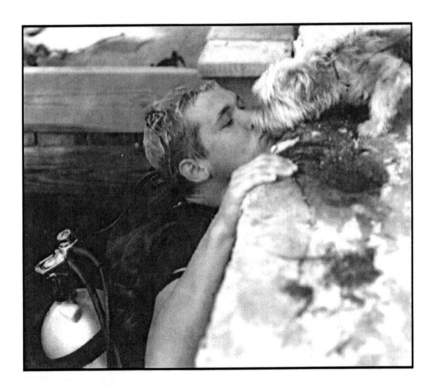

15 Year old Aquanaut Eric M. Chamberland
bids farewell to his dog, Alex, and departs
for a day of living undersea.

"We need a renaissance of wonder. We need to renew, in our hearts and in our souls, the deathless dream, the eternal poetry, the perennial sense that life is miracle and magic".

-- E. Merrill Root (1895-1973)

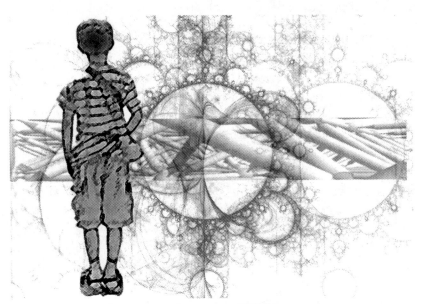

HANS IS DEAD

"Ah, great it is to believe the dream as we stand in youth by the starry stream; but a greater thing is to fight life through, and say at the end, 'The dream is true!'"
Edwin Markham

Many books prior to this one about human activities undersea have been written from a technical perspective and have blended both science and technology into a kind of hyper-technical reading. This work is intentionally not written from that perspective. I have deliberately gone through and attempted to scrub the white coat and pocket protector flavor, as well as the likeness of the wide-eyed and crazed Nazi Secret Society of Deep Sea Divers. Such an approach to undersea exploration was thoroughly described by record-holding diver Terrence Tysall in a mere six words — spoken with pointed finger (*at me*) and the proper accent, of course:

"Hans is dead. *You* dive now!"

Fortunately, Hans really is dead and we have no plans to revive him or appoint volunteers to replace him.

There is, however, one particular passage you will encounter near the end of this book that I want to pre-phrase up front so that you will understand my approach and the intentional essence with which all the words here were seasoned:

"This book openly discusses why undersea habitation by large numbers of people has not progressed, and many of those reasons are related to the fact that very few people wish to live in tiny cylinders and have to undergo risky and complicated decompression while talking like Donald Duck. Aunt Miriam does not even want to discuss taking her children (six year old Billie, eight year old Frank and five year old Misty) to live permanently in this kind of uncomfortable and often precarious environment.

In this book, I have taken the side of Aunt Miriam and the kids."

I have included as many figures and technical references as I could get by with, as any discussion of this sort demands before it can even be minimally understood. But I attempted to do so with the same respect given the average airline passenger who is not expected to scribble Bernoulli's equation on the back of a napkin to qualify for the red-eye to San Francisco.

As in any technical field dominated almost entirely by professionals and experts, those outside the 'community' are often shunned and even ridiculed for the unpardonable sin of lacking technical expertise. Further, there are still arguments aplenty within the community itself about what approach is best for long term habitation and there are, as in all professional communities, turf wars and bouts of dreadful jealousy that occasionally rear their ugly heads. All of this makes it most difficult to hold the door open for the novice or the newly interested individual just seeking information and testing the waters for themselves. Not only is the seeker at risk, but the poor doorman even more so!

I have been a member of the professional science and engineering communities all of my professional life. Admittedly, there is some snobbishness and the occasional vain attempt to claim kingship of whatever hill, but there is also a profound interest among most of my colleagues in engaging the public in what the science and technical community is doing. But please forgive them. They have sternly taught one another that science and its technical communication is only supposed to look and sound a certain way, or it is summarily dismissed, especially among themselves.

Thus, I have taken a risk by writing a book that seeks to engage Aunt Miriam and the kids, with a polite tip of my hat to my colleagues, all of for whom I have a great and abiding respect. Here you will find a genuine attempt at a deliberately personable approach to communicating a scientific and technical discussion, as well as a fictional tale at the end to complete the story.

After all, I have learned over the years that the principal task of the scientist or engineer is ultimately not to communicate with another scientist or another engineer. Instead, his most fundamental undertaking is to converse clearly with the public in terms they can comfortably understand and, if at all possible, to leave behind enthusiasm and passion for their work. If he cannot accomplish this most elemental task, for whatever reason, then his life's work is wasted and forever locked within his mortal self, regardless of the scope of his genius.

If the great undersea dominion encompassed by the single world-spanning ocean is ever to be populated *en masse*, it will certainly not be inhabited with the occasional wild-eyed and crazed diver, the eccentric engineer or the weird scientist (God bless them all), but it will be primarily populated by Aunt Miriam and her kids. If you want a glimpse of tomorrow's world undersea, go sit down with them, put your feet up, drop the jargon and ask a few pointed questions. Scientist and engineer types are not beyond learning new and astounding things themselves, even from the most unexpected sources. The answers are guaranteed to make any professional's head spin for days. Go ask 14 year old Ian Novak and his mom Shelley, whose answers we have recorded later in this book, and their responses will probably surprise you!

But thankfully, here between these covers, Hans and his Secret Society of Deep Sea Divers are dead and no one is really expected to take Hans' place. Now we can all relax and get on with the real discussion – and it is a most surprising and unexpected one. The whole concept of Aquatica and Aquaticans is stunning — but it is one seldom discussed or written about in even the most extreme science fiction tale. Yet today it is real and it is upon us. So, dear reader, go ahead and feel free to drop your preconceived notions about what a book of this kind is *supposed* to look and sound like, and sit back and enjoy what may prove to be the greatest human adventure in all of history!

Dennis Chamberland
Stonebrooke, Tennessee

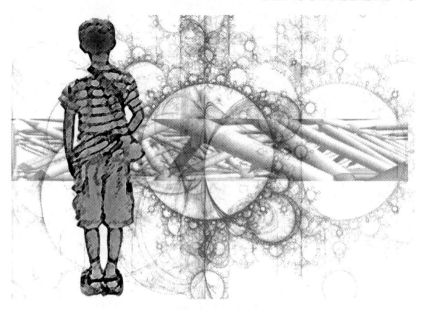

THE FIVE PRINCIPLES OF UNDERSEA COLONIZATION

First Principle

In every exploration system, we must require the systems we build to adapt to the human standard rather than expect the human to adapt to the machine or the environment – and in every design activity we will protect the human as a primary objective.

Second Principle

Every colony system will strive from the outset to minimize surface connectivity. Every possible surface connection will be re-deployable back to the colony if the need should arise.

Third Principle

Every colony system will endeavor to utilize *in situ* local resources to meet every need, from food to gas exchange to energy, in order to reduce dependence on land based resources.

Fourth Principle

Every colony system will set out to reduce their impact on the environment to zero, attempting, in as much as possible, to become a net positive environmental impact specifically because of its presence.

Fifth Principle

No wastes will be released into the ocean environment. Every undersea colony will become a zero waste generator to the ocean environment, freely providing a philosophical and technological model of this philosophy for other cultures.

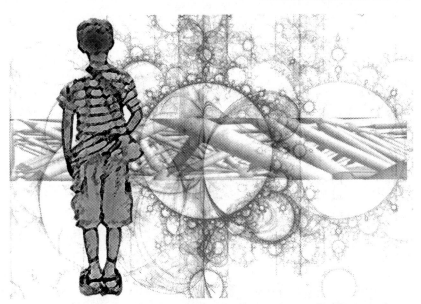

A SPLINTER IN MY MIND

"There are those who look at things the way they are, and ask why. I dream of things that never were, and ask why not".
Robert F. Kennedy

I was born with a dream, like a splinter in my mind. It will not go away and it has not faded, even over all these years. It has been my presumed destiny. But it is still unresolved – still hanging before me like a shimmering mirage I am forced to walk through or around each and every passing day.

The dream is living as a permanent human citizen under the sea. I have always hoped to be a true Aquatican – a permanent resident of Aquatica.

Even as I write, it is still an impossible dream – at least at this moment. It is impossible because of the 100 billion humans who have lived before me, and those who are living now, not one of them has built a colony in Aquatica where I may go live out any number of my days. I cannot pack my bags and walk to the shoreline of the North Atlantic Ocean, just a few feet from my door in Florida, and embark to live undersea in Aquatica. The population is, and has always been, zero. There is no place to go, no dwelling to go to and no fellow citizens of the great but empty empire.

Fortunately, I have met people and been to places that are very close to this dream since 1972 when, as a college student at Oklahoma State University, I initiated my original plan, that I called the Omega Project, to build mankind's first permanent undersea settlement.

In the intervening decades, as I have chased this dream all over the planet, I have been blessed with the friendship of Chris Olstad, an aquanaut with more undersea habitation hours than any other human being. Right now, Chris is as close to being a true Aquatican as any other human. I have also been blessed with the extraordinary opportunity to frequently visit and work out of the two longest-lived undersea habits in history – the *MarineLab* and the *Jules' Undersea Lodge* in Key Largo, Florida.

As well, I have been honored with the mentorship of history's most courageous and unsung undersea pioneer-hero, Ian Koblick, who has single-handedly kept the undersea regions open for the common man, and often at his own cost. Without the opportunities afforded to me by Ian Koblick, this book could not have been written.

I bring these individuals and circumstances to light because this quest has never been about what I have accomplished. It has been about standing on the shoulders of these giants who, by personal sacrifice and a commitment of their lives, have made all this possible. Further, they have also stood on the shoulders of others. I frequently remember Koblick's account of his meeting in the 1960's with Kennedy Space Center Director Kurt Debus.

Kurt Debus was on a quest to launch men to the moon. But when Ian Koblick walked through his door and asked for assistance to launch the inner-space research vehicle, the *La Chalupa* habitat, Mr. Debus agreed to help in any way he could. And so it was that all of us who enjoy access to the undersea world owe not just Ian Koblick, but Kurt Debus as well. And those are just a few pioneers who have paved the way to this point in history.

But here we are, dear reader. And if you will but open your eyes and look at me, you will see a man who stands at the shoreline; wife by my side, our bags are packed, our cat, Snickers, is tucked under one arm. All of us are ready to go – but with no possible destination.

It would be frustrating, except I passed the point of frustration a long time ago. Now I am just determined. It is going to be something I pursue right up to the end of my days. As I have told many, I will go there and live there and I will keep trying until they throw dirt on my face…whichever comes first.

However, having said that, this is not a book about me or my efforts, exploits or expeditions. Those books will come later. This book is all about where we are now, how we got here and where we are going, in the very near future!

Finally, I thought it was important that you know who it is writing this book:

I am not a man on a mission, but there is this mission embedded within my soul. This book is not all about me, but about the dream that has laid burning for decades like a splinter in my mind.

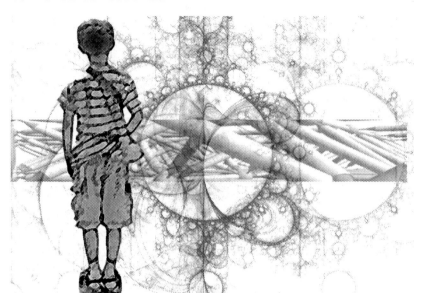

POPULATION: ZERO

"There has been a calculated risk in every stage of the American development.   The nation was built by men who took risks: pioneers who were not afraid of the wilderness, brave men who were not afraid of failure, scientists who were not afraid of truth, thinkers who were not afraid of progress and dreamers who were not afraid of action."   Brooks Atkinson

C ome with me to an altitude of 300 miles. Let us orbit our planet together. Look now down at our beautiful blue planet with new eyes. What do you see? You see primarily oceans.   In fact, the planet is practically covered with blue ocean water.

If you were an alien, you would almost certainly label Earth as 'the ocean planet' because of the predominance of its seas. The earth's seas are nearly everywhere; the land areas covering less than 30 percent of her surface.   From the orbital perspective vantage point, as we look down on the great Pacific basin covering over half the globe, except for a few islands, no significant land appears at all.

Said H.B. Stewart, "If in the past we have been prone to limit our geographical thinking to the land areas, we have been equally guilty of thinking of the ocean as something other than a single world-girdling sea. Such phrases as 'the seven seas' and 'the oceans' imply that, in addition to the seven continents, there are several separate and distinct oceans. The sea is, in fact, one global sea."

THE GREAT PACIFIC BASIN

As we continue to rotate above her and look down on our planet, we see the continents are teeming with humans – currently over six billion and rapidly increasing. Yet, most astonishingly, regardless of his self confidence, and for all of his technological prowess, mankind lives crowded upon, and fighting over, the minimal land area of the planet. Even so, the whole of his undersea dominion – except for the occasional subsea voyager – is astonishingly, totally, absolutely unoccupied. Here, the permanent human population is exactly zero.

What we see from orbit is a whole planet nearly unoccupied. And what space is occupied is a two dimensional surface of some 57 million square miles, packed with billions of people sometimes fighting and dying over a few square miles of dried up, overused, abused land. Yet, just beyond all that conflict lies more than 321 million *cubic* miles of area *with a population of zero*! As of this writing, not a single human has ever claimed the vast undersea regions as their permanent dwelling place, ever. Not only does no one live there now, no one ever has.

Why? How can this be?

That question has not only always haunted me, it remains the lingering, evocative mystery of the reality of human eccentricity. While we dream excessively about, plan for and spend untold billions on the vastly more difficult, risky and expensive establishment of space, lunar and planetary colonies at millions of miles distance away, there has never been a push to permanently inhabit the whole region of unoccupied undersea space just a few feet from us.

If we look at the expression of our curiosity, it is unthinkably astonishing to me that we, as humans, know far more about the resolution of at least half a dozen planets – including Venus and Mars – than we do about our own planet. Scientists can tell you with absolute certainty more about the details of these distant bodies than they can tell you about the details of the surface of our own planet beneath the seas. It is bizarre, but it is true.

Why? How can this be?

I must tell you that, having spent my professional career as an engineer and a scientist, I cannot put my finger on the reason. It is beyond my ability to understand and I simply file it in the category of human weirdness. Human beings are not only subject to bouts of individual oddness but, as a species, social oddness as well. It actually seems logically consistent to our culture that we are passionate students of distant planets but manifestly ignorant and wholly uncaring about our own. And this state of affairs seems absolutely reasonable to most of my fellow humans.

As we look at a vast region of earth whose population is exactly zero, we are eager to place blame for this human lapse of judgment and astonishing failure of collective creativity. Primarily we blame it on the lack of technology. The fact is, humans have had the technological capacity to populate the undersea region since the dawn of the industrial revolution. In 1942 humanity's last excuse was used up when Jacques-Yves Cousteau and Emile Gagnan invented the 'Aqua Lung' that allowed humans to breathe underwater. From 1942 to this date we have been wasting opportunity after opportunity to permanently settle more than 70 percent of the earth's habitable spaces.

Then we blame it on a 'reason' to live in these regions. If we can fight to the brink of human extinction over mere hundreds of square miles of dry land, does this not speak volumes to the need to open up new territory to those looking for somewhere else to live? If we can pay millions of dollars an acre for prime real estate, does that not perhaps indicate the need for a new land rush somewhere else?

Then we blame the lack of undersea settlement on 'environmental sensitivity'. The fact is, as we will cover in depth later in this book, we are grossly incompetent stewards of our environment for not being undersea watching, collecting information and protecting! It is almost as if we are excusing the damage that is piling up there because we do not know about it. If it is out of sight, then it is indeed out of mind. Inexplicably, we are far more knowledgeable about the environment of Mars than we are about our own undersea regions. Even though the very health of the earth's atmosphere and our land areas is directly related to the health of the ocean regions, we simply do not seem to care.

While the permanent population of the undersea regions of the earth is zero, there have been visitors, to be sure. Submariners (defined as persons moving about in undersea craft) ply this territory every day on missions of national defense. And there are a few researchers who report there temporarily on scientific missions. As of this writing, there are only three known relatively full-time undersea habitats in operation in the world to which different teams of aquanauts report for duty from time to time. Ironically, all three of them are located within 15 miles of one another in or near Key Largo, Florida.   They are, in order of longevity and logged hours of operation: the *MarineLab* habitat; *Jules' Undersea Lodge;* and NOAA's *Aquarius Underwater Laboratory.*

Other than these three, there are other habitats pressed into service now and then around the world on relatively short missions.   One such habitat is the *Scott Carpenter Space Analog Station* that I designed and built for missions in 1997 and 1998. Other examples are Morgan Wells' *Baylab* and Lloyd Godson's *BioSUB*, with its 12 day mission in early 2007.

And yet, while there are only three regularly used habitats, there are many times more spacecraft built and launched to explore off-earth regions.   While space exploration is not a bad thing, and is equally necessary for different reasons, it is most odd that we have virtually abandoned our own planet in terms of knowledge and settlement while we spend vast resources to understand and settle distant worlds.

Of all the humans who have lived since the dawn of civilization – about 100 billion of us – only the tiniest fraction have lived beneath the ocean in a fixed undersea structure for more than a day.   And of them, only a small fraction have spent a week or more living and working there.   There are many times more astronauts than aquanauts who have lived and worked beneath the sea for more than a week.

By some fortunate allowance of divine fate, I have lived and worked undersea for more than 30 days.

I also thrived there.

My wife and co-explorer, Claudia Chamberland, and I earned our aquanaut certificates on October 13-14, 1993. On that day, we became card carrying citizens of the undersea world. We conducted many other underwater research missions and then, during the summers of 1997 and 1998, I and my crew successfully launched and lived in the *Scott Carpenter Space Analog Station* undersea habitat on the floor of the ocean at Key Largo, Florida. It was without question the most stunningly interesting and captivating time of my life. Before I had the honor of that fantastic experience, I considered myself an undersea fanatic. But after that experience, I was no longer just a fanatic – I was indeed captured and held for life.

This book is not about that small journey, but it is about all that lies ahead. This is a book about the last great frontier on earth. It is about the surprising and seldom spoken of potential that lies just beyond the reach of the white froth of any nearby shore. This book is all about the empires yet to be built, the vast nation states of the undersea regions that will someday come to rival or even dominate those of the land. This work is a work of daring vision that is seldom spoken of or imagined among men. And, yet, it is as surprisingly achievable as it is inevitable.

As we embark on this journey, it is essential that we begin by wiping the slates clean. There have been more than a few misdirected starts in undersea exploration. There have been quite a few misconceptions communicated to the public. This book will attempt to dispel and redefine those as we explore together so that, as we view the future of manned undersea colonization, we may do so with a more realistic viewpoint. Hence, it is important that you and I begin by trying to set aside everything we have learned and been exposed to and start afresh in our approach. In that fashion, we can together look at the vast oceanic expanse with eyes that have gained knowledge and insight from the contributions of past explorers and apply those lessons to the next great push into the oceans; not as visitors this time, but as permanent citizens of a grand, new world.

The world of the fantastic ocean empire that has yet to unfold is startling and new; I dare say, it is totally unexpected and unanticipated even in science fiction literature. In truth, every single civilization, large or small, powerful or insignificant, began just that way. No one could possibly have guessed the powerful culture Christopher Columbus' magnificent discovery would create centuries later. Likewise, as we open up this vast, new, unclaimed territory, with that opening will come magnificent new colonies, cities, nations and empires.

How can I be so sure of that? Because, in all of human history, it has always been that way.   Every great push into unknown frontiers has resulted in a new and powerful human presence. Further, the ocean regions are so vast, mostly unclaimed and so full of natural resources and strategic potential, that human settlement and domination is, in fact, inevitable.

This work is aimed at discussing many widely varied ideas; from history to engineering, legal entanglements to nation building. But in such a wide, virtually unexplored region right at our very feet, you would inevitably expect complexity and even intrigue.

What has truly astounded me over the years is the quiet before the storm. As a Native American boy growing up on the Oklahoma prairies, I fully remember watching in the springtime the black, roiling clouds of approaching tornadic storms. I will never forget the calmness of the air which stood virtually motionless even as the wall of wind, hail and dreadful natural power approached. With eyes tightly shut, it was as if there was no storm; no approaching change that would threaten all we had ever known.

Today, there is a deafening silence before the storm of the advent of the new undersea empires. It is so deafening and so absolute that most deny there is even an imminent change hanging close aboard. And yet, it is there nonetheless, poised to begin at any moment. When it does, all that we have understood of our future will be utterly changed. Tomorrow will be far different – and dare I say, far more wonderful than we could have guessed in our wildest imaginings!

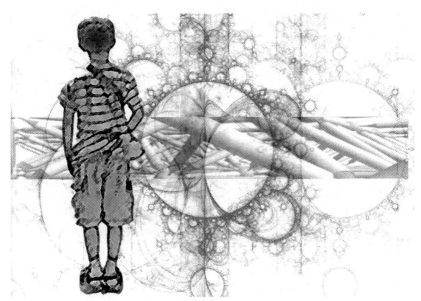

AQUATICA

"Greater than any army with banners is an
Idea whose time has come."
Victor Hugo

L ooking down at our mostly uninhabited planet, we see teeming billions of humans all crowded to her shorelines.   Yet not one of them has dared to permanently venture even a single step beyond those boundaries and under her watery immenseness.   It is equally curious that while this vast region lies all around the planet, no one has named it. The humans have named the dry areas all above her, but not the territory beneath the waves. It is as though this planetary ocean is deserving of only surface designations and is not considered for what it really is – a vast, connected, three dimensional whole. Earth is not a planet with seven seas; but instead we inhabit an ocean planet interrupted by an occasional land area.

If the ocean areas of the world were a single interconnected dry land, they would be called a 'super continent' and probably named Pangea.   But instead, paying attention only to the geo-political boundaries, the joined oceanic regions have been divided into seven blocks with predominantly unsystematic boundaries. In all of mankind's civilized history, the vast, three dimensional, wholly unoccupied regions of the earth have not even been named. Until now...

It will be called Aquatica,
and her permanent citizens will be called Aquaticans.

Aquatica is quite unlike the surface land areas that are wholly two dimensional; their citizens forever captured by gravity and forced to walk around on its surface. It is not so in Aquatica. Here the world is three dimensional and the Aquatican citizens are free to explore in all of these dimensions. They are always 'weightless' in their surrounding 'aquatic void.'

The world of the Aquatican is not bound by a surface. It is shaped not only by X and Y axis', but also by the added dimension of Z. The Aquatican habitation structure may be built on a flat undersea plain, but the Aquatican is also free to build in the center of their space, or near the top (also called 'the surface of the water' by the land-dwellers). Therefore, Aquatican colonies or cities may easily be built in suspended layers, one above the other. More practically speaking, Aquaticans are not limited to a single surface-bound location, but are free to construct their habitats and dwelling places in their vast cubic space for any reason at all!

The human presence in Aquatica, as described in this book, is 'Undersea Colonies'. And, although the fresh water regions of the earth are barely mentioned here, most of the principles herein would apply to them as well.

Exactly as in space exploration and habitation, Aquatica is mostly dark and cold. Aquaticans will have to engineer for pressure, temperature and exposure to extreme environments. The first Aquaticans will have to be engineers and technically savvy to build their habitation spaces safely and with user-friendly interfaces that will maximize movement and livability within their world. Yet, near the top of Aquatica, the space is illuminated by softly filtered sunlight dominated by blues and greens. Aquaticans may choose to exchange the relative peace and constant, undisturbed serenity of the dark regions below 100 meters for the better lit and more energetic layers above.

It is far too early to describe the world of Aquatica and how it will appear in the coming centuries. But Aquatica will become the next, and probably last, empire on earth. It will be populated by citizens, some of whom may live out their lives and never see the unfiltered light of the sun or breathe the air of the surface. Indeed these actions may not be considered freedoms at all, but liabilities and risks.

Aquaticans could possibly purposefully shun the light and energy of the surface. They may find strange the surface requirement to move about all day against the imposing force of gravity pulling ever against them rather than having the ability to float freely between one point and another. Some might wish to avoid crowds and the general cacophonous environment of the humans who live above, perpetually surrounded by air, dust, noise and crushing human masses.

Aquatica is a land wealthy beyond all current human assumptions. We are just beginning to discover that near the strange sub-marine vents called 'black smokers' lie deposits of metals and gold nodules in concentrations far more rich than any ever discovered on the dry land. Further, there is limitless energy to be harvested from waves, currents and swells that can easily supply all the needs of an undersea society in nearly any location on the planet. Without boundaries, Aquatica will not only dwarf any country, it will border all of them and thus, by definition, may become the greatest empire in human civilization.

The manifest destiny of humanity suddenly appears much different that we ever supposed, in a place right under our noses that, until now, has never even been named.

Today, as I write these words, it all sounds a little preposterous and science fiction-ish. By our own technological and strategic bungling, we have succeeded in ignoring and minimizing the final human empire – the greatest and most important of them all. But that day has ended. A few human eyes have finally opened and look to Aquatica and see it for what it is destined to be.

Many will soon follow…

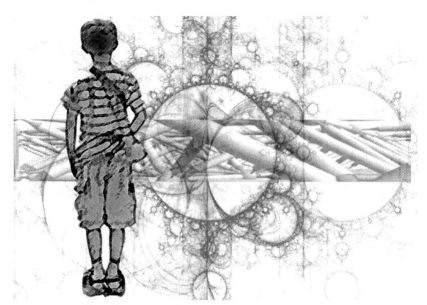

SOME IMPORTANT CLASSIFICATIONS

"It is never too late to give up your prejudices."
Henry David Thoreau

Many people find it surprising that mankind does not already permanently inhabit the world's oceans. Just as there was much ignorance and misinformation in the 15th century about what lay over the misty and indistinct horizon, there is much misinformation still, even in this age of millions of SCUBA divers and fleets of submarines. This lack of definition has led to many misunderstandings when discussing the permanent human colonization of the oceans.

For example, someone could rationally ask the question, "Why do you not consider it true that mankind has been a permanent occupant undersea since the first time that submarine fleets were launched?"

That line of rationale is about as accurate as stating that humans have been permanent residents of Interstate 95 since truckers have been driving up and down its length.   Transient undersea travelers have never been considered permanent residents, particularly since they have bases that are on land and they never stay underwater to go home at the end of each mission. And, of course, I-95 is not considered by anyone to be a human community.

Submarines and submariners are not at all the same as aquanauts and habitats.   Just as a 16 wheel semi-truck is not considered a human colony, a submarine is not considered a colony for the exact same reasons.   A truck and a submarine are moving vehicles and their passengers are obviously passengers – not residents of a highway.   Further, a truck driver is not considered a permanent resident of every city and town they drive through.   Obviously, they are considered transient drivers, not permanent residents!

Likewise, a submarine and an undersea habitat are also very different.   A submarine is a moving vessel whose designed purpose is to travel from one underwater point to another.   An undersea habitat is primarily a fixed structure whose purpose is to provide shelter and long term living accommodations for aquanauts.

A typical SCUBA diver is likewise not an 'aquanaut'.   An aquanaut is an individual who lives and works in a fixed habitat beneath the sea for more than 24 continuous hours without returning to the surface.   Most SCUBA divers return to the surface in less than an hour.   Further, they all go to their homes on land, where they actually live for the other 23 hours of that day, not beneath the ocean.

An undersea habitat is a fixed structure beneath the ocean that is designated for occupation by aquanauts who live and work there.   It can be fixed to the seafloor or tethered from the seafloor. Future designs may even have means of propelling themselves to various locations for harvesting or gathering operations, or for safety concerns due to geological or disturbance issues.   However, their primary function will be a fixed, permanent habitation for the Aquaticans who dwell within them.   A true undersea habitat is specifically designed to meet all needs of the human aquanaut – from comfort to shelter – for long periods of time.

A more formal definition of a human undersea habitat is: a fixed undersea structure that provides both working accommodations and full life support capability to allow for extended or permanent human occupation by a single human or team of humans for the function of carrying out the various full ranges of processes that allow the human to live undersea. The habitat provides the total spectrum of living accommodations for all human functions from eating, sleeping and waste accommodations to socialization requirements.

For the purposes of this work, let us now define undersea habitats in four categories:

The Platform Mounted Habitat: An undersea habitat may be mounted beneath the surface of the ocean on a platform that is fixed to the seafloor. An example of this would be a habitat that was mounted to the legs of an abandoned oil rig or other like structure. Such a habitat would have the advantageous position of using the existing structure as its ballast and its rigidity against storm and swells. It would also allow precise placement of depth in an artificial reef environment.

The Hybrid Habitat: These habitats consist of structures that are partially submerged and partially above water. They can be of two types:
One - anchored to the bottom but floating partially out of the water. This habitat definitely enters the 'gray area' of habitat definitions, since, if it has the capacity to move, it is no longer a habitat but a kind of 'glass bottom boat'. And, if it can be towed, it becomes a 'glass bottom barge' regardless of how well equipped and apportioned it is;
Two – a structure that is permanently fixed to the bottom, but provides living space both above and under the surface. Obviously both of these classes would normally be pressurized to one atmosphere and circulate air directly from the surface.

The One Atmosphere Resort Habitat: As of this writing, several of these one atmosphere habitats have been planned around the world, although none of them have actually been constructed (they are exceedingly expensive.) These habitats enable guests to 'walk' into the habitat by tunnels, stairs and elevators. The air pressure is the same as surface pressure and they are generally ventilated from the surface. They offer undersea views in spectacular and well appointed rooms. Generally speaking, they are bottom mounted structures that rise to the surface to enable easy access, ventilation and full life support from circulated surface air.

The Blue Water Undersea Habitat: The true Blue Water Undersea Habitat is a habitat that is totally isolated from the surface except by cables carrying power, breathing gasses, water and communications. It is only accessible by direct diving or undersea transport. It creates its own pressure and atmosphere. It is a structure whose purpose is much like a land based home. It is specifically designed to indefinitely meet all aspects of life and living beneath the surface of the sea and, except for its small cluster of support lines called an 'umbilicus', it is always completely isolated from the surface. Of these, there are three distinct classes:

Ambient Pressurized: These habitats are pressurized to the same pressure as the outside water at the moonpool hatch (the open hatch at the bottom of an ambient pressure habitat where water is held out by the pressure of the air inside which is equivalent to the pressure of the water at the opening). Hence, there is no 'closed door' to the outside, since the air pressure holds the water out.

Pressurized Less Than Ambient: These habitats are not pressurized to depth, but their interior pressure can be set to whatever depth the designer decides. This generally allows aquanauts to return to the surface without decompression, but does not allow immediate access to the sea without first passing through a lock-out chamber.

Hybrid Pressurized: This habitat is compartmentally pressurized by work space function. One part of the habitat may be pressurized to one atmosphere while another part is pressurized to depth. This allows for a large moonpool and workspace to be maintained while the aquanauts live in the one atmosphere space.

At this point I must mention the Engineering Cofferdam which is sometimes called a 'welding habitat'. These are the crudest types of undersea structures and there is some significant argument as to whether they should even be included in the definition of an undersea habitat at all. Their sole purpose is to create a dry space underwater to allow a construction or repair process to be worked upon in a dry environment. After the job, the cofferdam is disassembled or laid up until it is required for another task. I do not feel there is any justification for including cofferdams in the definition of a true undersea habitat since their design purpose was never 'habitation' per se, but providing a crude workspace, and they are not at all designed for 'living' undersea. For the purposes of this work, cofferdams are not considered 'undersea habitats' and thus they will not be included in any further discussion.

This book is focused entirely on the Blue Water Undersea Habitat in all of its different classes. While One Atmosphere Resort Habitats are certainly the wave of the near future in exposing many potential permanent residents of Aquatica to their first luxurious taste of undersea living, they do not represent the future of permanent undersea habitation any more than the Hilton represents suburbia.

The additional classification of depth needs to be presented as well, regardless of the structure's category:

The Shallow Water Habitat: Habitats whose hatch depth is equal to or less than 21 feet of water. This is the known limit of human decompression. If the aquanaut breathes an air-normal atmosphere at this depth/pressure, then no decompression is required to return to the surface.

The Deep Water Habitat: Habitats whose hatch depth is greater than 21 feet of water. Aquanauts breathing gasses at or below this pressure must decompress before returning to the surface. While it is true that hybrid habitats may be deeper than 21 feet and feature living areas at a pressure less than this depth, it cannot remain an ambient water habitat without decompression or engineering of a hybrid component to shield the aquanauts from a pressure greater than 21 feet – hence its designation as a deep water habitat.

In this work, the underwater/undersea regions of the earth are collectively called Aquatica and the permanent undersea dweller is called an Aquatican.

These definitions follow a well defined historic pathway. Explorers of the 15th century became 'New World Explorers' only after it was determined that there was, in fact, a new world and the intrepid few actually planned expeditions, pulled up anchor and sailed to it. Then there were a few New World Explorers who brought their families and decided to stay and never return to Europe. They became the first 'Americans', called after the land on which they settled. Eventually, the term American took on political and national meaning as it evolved from the name of a place to a sovereign government.

These classifications are important in order to distinguish between the true Aquatican Aquanaut, a citizen and resident of the sea, as opposed to the always transient submariner or curious surface dweller who descends through a tunnel to peer out some glassed-in chamber, but is not at all isolated from the surface, its air, pressure or effects.

An aquanaut is one who lives in a habitat for extended periods of days, weeks, months, years or decades without surfacing – but has plans to return to the land areas above Aquatica. An Aquatican Aquanaut is one who has moved to Aquatica with permanence in mind.

It is certain that these classifications will elicit some level of controversy, much like the debate engendered when German cartographer Martin Waldseem named the new worlds west of the European continent 'America' after the exploits of Italian merchant-explorer Amerigo Vaspucci and not after the explorer who had first discovered the land. Even as the author of this work, it is by no means certain to me that these terms will hold true or survive the ravages of time, personality and debate. But it matters little. In an unsettled territory many times larger than the small toeholds humankind has managed to eek out on their planet, there is no one there to make their claim; no precedents, no guideposts and no guidelines.[1]

Yet, in the interests of going on with this work about the complete unknown, I thought it simple polite prudence to give the places a name so that the discussion could at least proceed.

No more *Mare Incognito* on the maps of this book.

---

[1]This decision remained controversial, even though Columbus himself could not have argued against it since he went to his death convinced that he had, in fact, discovered colonized islands of eastern India, not the islands of a whole new continent that lay nearly half a world between them. It has been said, with some degree of cheeky accuracy, that Columbus was the only explorer in history who did not know where he was going, did not know where he was when he got there, and did not know where he had been when he got back!

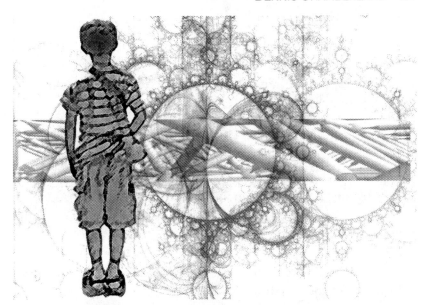

ONE DAY IN THE LIFE
OF AN AQUANAUT

"I am only one. But I am one!
I cannot do everything, but I can do something;
and what I do, and can do,
by the grace of God, I will do."
Anonymous

I have lived in habitats beneath the sea for weeks. Some recollections are stronger than others, but the thing that I remember most profoundly is the astonishing peace of the experience as I awoke to begin each new day. Living in a habitat beneath the sea is very much like living in a cocoon; protected against an alien world for which we were not created, and yet permitted to be there by that very cocoon.

As I awoke each morning, I clearly remember the sounds of the sea and the habitat blending together in a kind of otherworldly harmony. I would lie in my bed and, even before I opened my eyes, begin to assimilate the sounds, smells and essence of the day that was about to unfold. There is something quite different about waking up beneath the surface of the sea. As soon as your mind fixes on the reality of where you are, the mental processes begin to focus on the sounds of the machines that keep you alive in a region of the planet where there is no air and the environment just inches away is so inimical to life and the air-breathers that have dared to come here to dwell.

My mind sensed that outside and above were thousands of tons of seawater held at bay by the very air that I was breathing. Somehow, through various technological miracles, the air itself had become as powerful as the sea and literally forced the water to stay outside the always open moonpool just feet from where I slept.

It seemed odd to be able to stand up and look at the moonpool. There the water lay quietly at the air's interface, exactly defined by the shape of the hatch. On one side was the shirt-sleeves, air conditioned atmospheric mixture we breathed and that kept us totally comfortable. On the other side of the moonpool lay the Atlantic Ocean – all of it – from Florida across to Europe and Africa. I joked many times that the aquanaut has the largest swimming pool in the entire world at his very doorstep – the whole ocean itself!

The moonpool also doubles as a large aquarium. You can sit quietly at its edge and watch fish gather around and swim lazily underneath. As far as I know, no huge tentacle has ever reached inside and grabbed an aquanaut for dinner. Nor has a killer shark ever leapt up and snatched a tasty human for a midnight snack – except in the movies. But then, we have not been living in the oceans for very long!

As you sit and contemplate this always open door to the ocean in an ambient pressure habitat, it is likely that some creature that lives in this wide ocean should eventually stick its head up inside and say hello. As a matter of fact, it is not only possible, but it happens frequently. At night, small fish are attracted by the lights of the interior of the habitat. Larger fish are attracted by the smaller fish. Eventually, push comes to shove and, inevitably, some fish leaps out of the moonpool and up onto the deck. It is the aquanaut's sad duty to collect these unfortunates and toss them back as food for the larger fish outside each morning. One aquanaut I knew actually laid them onto a cracker and ate them! Waste not – want not. Breakfast is breakfast.

But as I lay in that half-sleep, half-wakeful state each morning, my mind separated out the sounds. In a house on the land there are sounds, too. There are air conditioning and heater sounds; sounds of gas passing into a stove burner from a regulator; sounds of water running in a water line; sounds of fans and the faintest hints of air whistling from outside to inside and inside out. But the sounds of an undersea habitat are altogether different. And each of them is far more immediately important.

If the power fails in a land house, the sleeper sighs and falls back to sleep, if he even realizes that it happened at all. The sleeper on land has the vast luxury to pile on more blankets – or kick them off – and wait it out. But undersea, the situation is far different and more immediate. Each sound represents one little part of life support. And without each part, life is threatened at once.

In our ambient pressured habitat, the carbon dioxide that we breathed was diluted and mixed by the same fans that blew cool and dehumidified air from the onboard air conditioner. This single machine was doing three vital things all at once: it was mixing the carbon dioxide evenly; cooling the air; and dehumidifying the air. Without all three, we would not have been able to live there very long. The sound of the constantly running fan on the air conditioner was not only a soothing white noise, it was also the sound of life and security. If that resonance had ever died away, we would have been counting the minutes until we would have had to retreat back to the surface. The habitat's air conditioner was literally one sound of life.

Early on in the first days of living and working in the sea, it was discovered that in the tight confines of a habitat, humidity quickly becomes a problem. Without constant humidity control, habitats rapidly become unbearable. It makes common sense, of course. Aquanauts are frequently in and out of the water, taking showers and hanging up wet clothing. Further, the air is constantly laced with added humidity from human metabolism through their exhalations as well as their perspiration. A habitat is the perfect water vapor generator and without control it makes underwater living either seriously uncomfortable or altogether unbearable. But the same air conditioning technology used to cool and dehumidify land based homes can be employed in underwater habitats as well.

In our habitat, we also heard the constant dull rumble, and even felt the vibration, of the air pushing past the moonpool's specially designed air-siphon. Instead of installing tricky and expensive carbon dioxide scrubbers, we had the luxury of using a shore mounted air compressor to provide fresh air in such quantities that it diluted our waste gas – carbon dioxide – to the degree that it was never a problem. That dull rumble meant to us that oxygen was being supplied and carbon dioxide was being pushed out the hatch. If that sound and gentle pulsation ever stopped, it would have been time to start thinking about packing up and leaving.

Some habitats feature air scrubbers. In these habitats, the sound of the carbon dioxide scrubber and the gentle hiss of added air/oxygen make up for the huge, shore based compressor that we used. But each noise represents the same effect: life being supplied to the aquanaut, enabling him to live and work beneath the surface of the ocean.

All these sounds join together as the aquanaut is curled up in their blanket. The air conditioner typically makes the air cool at night, a byproduct of squeezing the moisture out of the air. It feels very good to curl up in the blankets during the early hours of the morning and just lay quietly. There the sounds and vibrations blend together as one. The aquanaut lies silently, hearing the noises and feeling the gentle trembling of the great mechanical undersea creature whose metal organs all work together in a soft, synchronized vibration, all designed specifically to protect and preserve human life under the sea.

As I lay each undersea morning, eyes half open, absorbing the first few moments of the day, curled in the soft warmth of my bed, the gentle vibrations of the machines and escaping air seemed to me to be contrived to encourage rest and the deepest, sweetest sleep I had ever had. In such a cushion – layered with steel and insulated by soft cotton – it was most difficult to move into wakefulness.

The mornings were also laced with the hum of the refrigerator compressor kicking in and out, the occasional ill defined gurgles and drips, and the slow breathing of other nearby aquanauts. In all, the sounds of the new day were gentle and reassuring and reinforced the peace of the extraordinary place we offhandedly called our habitat.

Morning dawns beneath the sea with a far different tempo than onshore. The light of the undersea dawn is always a muted affair and comes about with a deliberate slowness. No matter what the state of the brilliant topside day-break, the undersea morn is never, ever harsh and could never be accused of being in a hurry. On the seafloor, the reds and oranges of sunrise are obliterated. The morning dawns as an advancing gray light that does not peak as the sunshine does on the surface. It intensifies slowly, softly, as a gentle illumination that never climbs quickly – but it is one that changes in its sweet time from subdued grays to dull blues, then mounts to a green glow.

I made it my daily habit to awaken and sit quietly at the moonpool with my feet in the water, watching the fish warily gather around my toes, their wide eyes looking up to me for their morning handout. They knew me too well, it seems! Each and every morning we played out our ritual. I would stare at them unblinking, silently and still. I waited patiently for them to inch their way closer and closer to my toes. Then suddenly, when they were but inches away, I would wave my hands wildly at them and growl and watch them scatter in half an instant out of sight.

Meanwhile, from her warm bed, Claudia would always ask, "Are you okay?"

"Yes, dear. Just playing with the fish."

Then they would slowly return, one eye on the crazy biped – his body mostly out of the water but his giant feet poised right before them. As they returned, I would break up a saltine cracker and dribble it into the moonpool as a reward for their endurance of me and the daily rights of human-to-fish morning terror. They returned day to day, because they either had a very short memory or they patiently tolerated the feral human.

Thereafter, I would slip a weight belt around my middle, attach a backup air source to my belt along with the long hookah hose, slip my mask over my eyes and slowly lower myself into the Atlantic Ocean which lay right at my feet.

No matter how warm the water was in reality, when the liquid flowed over my body, which had been warmed for hours by my blankets, there was a moment of mild dismay and I would suck long and hard on the regulator, forcing the air to rush into my lungs with a metallic hiss. But it only took a few seconds for my capillaries to quit complaining from their rude awakening and for the momentary discomfort to pass. From there I released my grip on the habitat and slid quietly out to the seafloor.

As I looked out into the ocean, it was never yet fully as bright as it would be at mid day. On the surface, morning had spread its light fully; but here, undersea, the light would slowly build for at least another hour. It is a trick of perception, of course. The light would also build for another hour on the surface, but there the human eye would have reached its saturation point by now. The pupil would begin to limit the light entering the eye, so that it appeared to be fully bright, even though there was another hour of increasing intensity to follow. Yet here, 30 feet beneath the surface, the eye was still waiting for full illumination and things would seem to brighten even more as the morning light matured.

I usually swam about a fathom from the moonpool, turned around and sat on the seafloor looking back at the habitat that rose above me. It was always an incredible sight. It was a structure of my own creation, first hatched in my mind, then crudely drawn on paper, turned into engineering drawings, cut out of steel plates, welded together, outfitted and tested, launched into the ocean, purposefully sunk and now it had been my home for these many days and weeks. I could not look at it enough from this vantage point on the ocean floor every dawn in the diffuse morning light.

It was not just a dream come true, but it was a kind of miracle as well. It was a house for living and working undersea – for six weeks each summer, it was my address. On all the planet on that day, there were less than half a dozen of us living in habitats on the ocean floor. And on that same day, more people were living in space than in habitats under the world's oceans! Perhaps six people out of seven billion called our habitats and the seafloor home for that day. And, sadly, all of us were just visitors. Not one had come to stay. In yet another week, this population of humans would rotate to a wholly new handful, all coming down for brief stays.

I loved to sit in the gathering light and look back at my undersea dwelling. While I sat at 30 feet on the bottom, the moonpool was above me at the 21 foot level – the absolute limit for non-decompression saturation. My blood was saturated with nitrogen. If our moonpool were but a single foot deeper, statistics would not be favorable for surfacing without a prolonged decompression.

The fish I had just fed from the comfort of the wetroom now kept their distance but circled before me nonetheless; still curious to see if the strange giant would feed them from this position. I tried never to disappoint my finned friends and sprinkled before me, as best I could, some soggy cracker crumbs. They always responded in their wild ways by darting in for the capture, then literally turning tail and shooting away. While I did not have them on my mind for breakfast, many other creatures here did, so such skittish behavior in the deep was not considered socially loutish, but instead, quite wise.

I would watch from my external vantage point as the ever present curtain of a billion and one bubbles escaped from the edge of our air siphon and ballooned toward the surface, carrying with it our carbon dioxide. The wall of bubbles silently skimmed past the sides of the habitat while in and out of it darted tiny minnows called the Atlantic Silverside. These small fish were always present and usually hung out in the bubble curtain. Since they were near the very bottom of the vertebrate food chain in these parts, they appeared to know this well and behaved accordingly. They were as fast as lightning and traveled in well organized schools.

I loved sitting quietly on the bottom of the sea, watching the myriad life forms prepare for the daylight hours. Many organisms preferred the relative safety of the night hours for their outings but, if you were very careful and looked closely enough, it was easy to see the nocturnal creatures digging their daylight holes into the bottom or hiding in the tiniest nooks and crevasses of the habitat's structure as the undersea day brightened.

In the open ocean a safe haven is always at a premium. Some marine biologists call this 'niche surface area'. In the open ocean, niches are hard to find, and where there are no niches, the population cannot settle in and grow. That is exactly the premise behind artificial reefs. Any new niches will do: old tires; open concrete blocks; sunken vessels – all provide an increase of niche surface area for new species to make a home and thrive. Now I had built for them a new reef. It was my human habitat; and *my* niche was carved out inside of it.

As promised, I built it and they came – in droves and in all sizes. The Silversides found protection in the bubble curtains. Crabs found a new home beneath the pads of the habitat's legs. Larger fish took up residence in the folds and creases of the steel structure, backing into place and facing outward, ready to defend against any interloper that would dare challenge their territory.

Over many days, I came to recognize these creatures and their homes. I named the larger fish and a few crabs and, though I could not speak to them with a regulator sticking out of my mouth, I communicated with them using hand signals, which they all consistently seemed to interpret as hostile and raised their dorsal spines to prove it. It seemed that no matter how fascinated I was with them and their habits, they returned my curiosity with a clear sign for me to move on.

As on the shores of earth, some days in the water are better than others. Under the sea, there are 'cloudy days' when the weather topside causes silt and other particulates to reduce visibility to a kind of aquatic fog. On those mornings, excursions away from the habitat were either useless or not as enjoyable as sunlit days. But on a still and beautifully clear day undersea, the aquanaut can settle back and look up, watching their bubbles tumble silently upward and explode on the surface like a boiling pot.

As much as I deeply admired the close-in view of undersea life, I was also equally fascinated by the larger sight of the seafloor: the water column as well as the surface features as seen from the bottom. Here the view was one of otherworldly beauty. If the aquanaut's depth is right, the surface and its ripples, or even its mirror smoothness, can be discerned. At a distance, fish can be seen plying the empty aquatic space between the bottom and the surface. Sunrays can be viewed dancing like nearby aurora borealis. If the wind driven ripples and sunlight are just right, these rays can be fantastically detailed and reach downward with astonishingly beautiful depth and kaleidoscopic resolution.

While rainy and windy days above can often obscure the undersea visibility due to a variety of interactive effects, if at the right depth when the rain just begins, you can see the dots of the individual drops pepper the surface and watch it in all of its indistinctly silver impressionistic forms. You can hear the rain splash on the surface as a kind of distant white noise. At night, you can also see flashes of surface lighting as sheets of radiance from an unknown direction or origin.

Soon enough, I would always inch my way back to the moonpool. Watching all the aquatic life feasting on their morsels in the new morning always made me hungry. I approached the moonpool slowly, knowing that in so doing, I was going to aggravate the Bluestriped Grunt that kept his home in the moonpool trunk. As I approached, he invariably responded as he did on each and every one of my trips in and out of the habitat – he would raise his dorsal fins, open his mouth wide in a pair of cavernous gulps, then scurry quickly away. After countless trips in and out of the habitat by numerous people every day, I figured that sooner or later he would go away and stay away. But, up to the very last day the habitat was in the water, he remained at his post faithfully guarding his adopted space.

I would slide back into the habitat and, once again, feel the discomfort of passing from one medium to another. Each and every time I came up into the moonpool and broke the air-water interface, I was cold. The blast of air conditioned habitat air was even sharper than was the early morning slide into the water a half hour before.

I would pull myself up, sit on the side of the moonpool, strip off my mask and shed my gear, belt and hookah hose, attaching it to the clip on the wall. If left on their own, these 100 foot hoses attached to the air compressor ashore had a bad habit of spontaneously slipping away out of the moonpool and drifting to the surface. On more than one occasion aquanauts have been forced to call on surface crew to fetch the black, slippery, rubber snakes with hissing stainless steel heads and swim them back down.

Next, I would slip out of my suit and stand under the warm shower, wash the saltwater off of my body and feel the just-right sensation of my surface capillaries opening up again to the normal air interface. After quickly drying off, I would wrap my towel around me and sniff the delicious smells of the breakfast Claudia was preparing in the microwave.

Aquanauts are perpetually wet people, almost by definition. But well designed habitats are most definitely not wet places! So our habitat was designed to keep the wet areas wet and the dry areas dry. The wet room was the place where the moonpool was located as well as the shower. There was also a nylon barrier that separated the wet room from the other part of the habitat. In that way, the dehumidifier/air conditioner could spend its time scrubbing the dampness from the living area while at the same time no one really cared whether the wet room was... well – *wet*! Long experience has proven that living 24/7 in a damp environment is not only no fun, it is downright miserable. That led to a set of aquanaut–habitat rules.

I originally thought it would be impractical to ask visiting aquanauts and technicians to shower and change into dry clothing on each and every entry into the habitat. So our rule was: after receiving permission from surface Mission Control and the Mission Commander in the habitat, you may enter the wetroom through the moonpool; then decide to shower or not; dry off with your own towel; wrap it around your suit and only then could you enter the dry compartments. No wetness was permitted forward of the barrier, save for the exception of a damp suit under the towel.

Of course, bathing suit type immediately enters the picture. Bathing suit styles differ from year to year and even generation to generation and gender to gender. When the long term habitat projects began at Marine Resources Development Foundation (MRDF) in Key Largo, the style was Speedo suits. That was fortunate because the tiny nylon bikini Speedo does not hold much liquid, dries very quickly and is quite comfortable. The aquanaut staff at MRDF had worn these suits all day every day for years. By the time we arrived in the late 1990s, the fashion style had shifted radically to the oversized, baggy, and consequently very drippy and very wet board shorts.   Not being an individual who particularly cared about the latest swim suit styles, I adopted the Speedo of the aquanaut staff and never regretted it one minute.

As Mission Commander and 'owner' of the habitat, I was particularly annoyed when visitors arrived with the baggy, dripping, oversized suits that made a general mess. The lesson for me was that in future expeditions we will require dry clothing one way or another in the dry areas – even if it means visitors must pack dry clothing to change into upon arrival. Personally, I will always prefer the time-tested bikini Speedo and t-shirt. Besides, it is impossible to get any more comfortable than that! It is, to me, the official 'aquanaut's uniform'.

When I entered the habitat after my morning swim, I could immediately see that Claudia had switched on all the lights in the main cabin. A future chapter will discuss lights and illumination strategies, but suffice it to say that the human mind is wired for the proper light in the correct space. When our habitat was designed, we installed 'natural light spectrum' bulbs in the main cabin (with the advice of our world class – and probably only – undersea habitat electrician, Darren Corriden) which leant a pleasing and home-like yellowish light. The light was bright, but not too bright and made the space appear comforting. It was the perfect choice. Meanwhile, the light in the wet room was pure, bright white – also a perfect choice for that particular space. More about lighting in undersea habitats will follow later.

As I approached the interior space I would hear the voice of the surface Mission Control making their daily wake-up call. Being a 'morning person', I invariably awoke and was moving about at least an hour before official 'wake up'. I would sit in the commander's chair and chat over the radio with the sleepy voice of the command center operations watch. I would then make a call to our Chief Engineer, since I acted as his morning alarm, and we would discuss the engineering activities for that day.

While metal, the habitat was nothing less than a living beast. She was made up of constantly humming machines, circulating fluids and air movers much akin to a living organism. But unlike a living organism, she could not repair herself, and if anything went wrong, the habitat engineer and I were her physicians and surgeons. In a habitat on a prolonged mission, your system gives you two choices: have a backup system to replace the one that failed; or pray you can fix mechanical problems faster than carbon dioxide and/or temperature and humidity build up and chase you back to the surface.

ENGINEER AND AQUANAUT
JOSEPH M. BISHOP

Our habitat engineer, Joseph M. Bishop, was about as close to a Star Trek's Engineer 'Scotty' than any other living human. 'Serious' is not a strong enough word to describe Joseph's attitude toward his metallic baby. He had just about as many hours invested in the design of the habitat as I did. My design was marine architecture, shape, function and operations. Bishop's design was one of systems and functional integration. Joseph, a licensed marine engineer, nuclear submarine engineer, Merchant Marine Officer, hard hat diver and a graduate of Kings Point Merchant Marine Academy, is a large man with deep roots in West Virginia. Hence, it was never a good idea to call his baby ugly – even by mistake. Joseph and I would have a daily discussion of the habitat and her systems each morning before breakfast.

Before the mission began, we had purchased our entire mission's meals and put them in a large freezer topside. Each day, the duty aquanaut from the surface would bring our meals down in a watertight transport case. We would then install them into our small habitat freezer and refrigerator. When it came time to eat, we would slide the frozen entree out, pop it into the microwave and set the timer.

In our habitat the atmospheric pressure was 1.6 ATM – sixty percent more than surface pressure. That ever-present knowledge was of critical importance when operating the microwave oven. On the surface the temperature for boiling water is a well known 212 degrees Fahrenheit. But at 1.6 atmospheres, that temperature is 239 degrees. 27 degrees does not seem like much of an increase, but it is significant enough to burn skin easier and ruin food faster!

Meal time in an undersea habitat takes on a whole new dimension. Many astronauts tell the same tale about food in space. For some reason, food becomes very important and meal time is a real occasion to look forward to. I have given this much thought and I believe it is due to a level of real, but hidden, subliminal stress. While my psyche always loved being undersea and thrived there, I also believe that I was also engaged in some level of unconscious stress in this remote and extreme environmental setting. I have hypothesized that after weeks, or perhaps months, that subliminal stress would almost wholly disappear and mealtime might take on its old, traditional sense of duty.

Claudia and I would sit together and eat our breakfast facing the wide windows. Our habitat was designed so that as we sat and looked out, we could not only see in front of us, but also to the surface and to each side. It was an extraordinary sight to sit in comfortable seats, in a well lit habitat with a perfectly balanced atmosphere and get the wild ocean view while enjoying a hot meal. I never got enough of that view at breakfast – of the greens and blues of the ocean panorama completely surrounding us. It is true that there are some spectacular restaurants at which I have dined that feature huge windows into magnificent aquariums – such as Disney Epcot's Living Seas and the unparalleled glass wall looking out at the bay at the Monterey Bay Aquarium. But even they hold no comparison to the view from the seafloor, surrounded by the ocean's grand vista and populated by its unencumbered and utterly free inhabitants.

After breakfast, I would accept the chore of cleaning and straitening the interior of the habitat and collecting our daily trash for the duty aquanaut. Meanwhile, Claudia would get suited up to take her 'morning walk' outside and to do some external cleaning of her own.

The habitat was intentionally designed with wide and spacious windows. I had visited many habitats before, and I intended to fix what I felt were their two major drawbacks: lack of windows and comfortable seats. In a cylindrical habitat (all of them built for good engineering reasons) windows are placed on the flat ends. I have sat for countless hours at these, straining my neck and eyes to try and get some glimpse of the surface. It was never possible and the action of trying to look up and focus though the vagaries of a flat pane and the glass-water interface always left me with a tight neck and eye strain. And so it was that while I had spent many days in undersea habitats of various designs, it was not until my design sat safely on the seafloor that I was finally able to lean back in a comfortable chair and actually look up and see the surface ripples and light rays above me.

Unfortunately, with a lot of windows, there is also a lot of daily work that accompanies them. In the bay ecology where we were situated, there is a daily shifting and circulating of tiny particulates that float in the current. Likewise, there are countless millions of living organisms in the same current – many of them feeding on the smaller particles or on one another, and some just taking a fateful ride to wherever the current would carry them.

When the particles fell out onto the habitat – both living and non-living – they formed a light, daily blanket or dusting of material that, if left unattended, would quickly obscure our view. Further, in the fine drifting mix, there were also organisms looking for a permanent home that would attach themselves and begin to grow. Some of these were undersea plants and others were creatures such as the lowly, but entirely entertaining and quite fascinating, barnacle.

Claudia never swam outside without her thin diving suit called 'skins'. They were made of a fine Lycra layering that gave her modest protection from chill and the particles she would be dusting off the habitat's windows and skin. She would swim up to the front window of the habitat and begin there. Cleaning a habitat's windows is relatively straightforward: you simply wave your hand slowly an inch or so above the surface, and the particles float away in the blustery but momentary current. After that, Claudia would examine the windows for any attaching organisms and remove them. We would leave a few organisms around the edges that were in easy-to-view positions from the inside so we could watch them day by day without obstructing our view.

Barnacles were especially fascinating. They are organisms that form a hard shell over their soft interior bodies. They are well known for these hard and razor sharp shells. But if an aquanaut will get close from inside the window and look at their actions, it is more than interesting to watch how they feed themselves. From the top center of their hard shell, a set of what are called 'cirri limbs' (which to me look exactly like bird feathers) begin to sweep the water rhythmically before them. The action causes the particles that drift in the stream to become entrained in the cirri limbs and pulled into the barnacle's interior for digestion. I have watched these creatures with total fascination for hours as they endlessly feed themselves with whatever happens to float by.

We were also high enough in the water column so that the white top of the habitat particularly liked to collect different varieties of adhered sea grasses. Claudia did not spend much time on cleaning off the sea grasses, so that by the end of our six week missions, we had collected quite a growth of these ocean dwelling plants. If we had been in the water more than six weeks, we would have concentrated on keeping the habitat's windows clean, but we would have also encouraged the ocean to take over the rest of the habitat's steel skin and do with it what it wanted.

There was not an issue of rust, even though the habitat was made of steel and submerged in the bottom of the ocean. The pictures of the Titanic always come to mind when I think of steel structures on the seafloor. The Titanic's grand superstructure was literally dissolving, its once magnificent walls rusting away in gigantic rust-sicles and being absorbed into the sea. But that is not at all the only possible fate for undersea steel structures.

Rust forming on steel in the ocean is caused by an oxidative process akin to spontaneous electrolysis. Electrical current that naturally occurs in the ocean by various processes, including the complex chemical nature of the water, causes the steel to oxidize and it is consumed by the seawater one molecular layer at a time. But mariners found out long ago that if a metal that oxidizes easier than steel is attached, the current will be drawn to it and the steel will be spared. Such a metal, they found, is zinc. By bolting these 'sacrificial zincs' to the hull of the habitat, they oxidize instead of the habitat's steel! Aquanauts can thereby replace these sacrificial zincs every so often and spare any rusting of the hull of the habitat at all.

Just a few meters from our habitat was another historic habitat, *MarineLab* (which today holds the record for the longest manned habitat in service) that had been submerged in that same bay for over two decades, and its steel was just as pristine as the day it was launched. On the hull of that habitat grew a layer of living ocean organisms that had built a community in exactly the same shape of the habitat's skin, but there were no rust issues.

On many occasions, the top of the *MarineLab* habitat became my safety decompression station on my way back from a long stay undersea. I would cling to one of the support structures atop *MarineLab* and wait for the nitrogen gas to slowly and safely evolve from my body. Looking over its surface from a foot away was akin to looking down at Los Angeles from 35,000 feet. Before me was a living carpet of countless billions of organisms clinging to the surface of the habitat from the size of the tip of my little finger to the microscopic. It was a blend of plant and crustacean life of all shapes and configurations.

When I lowered my mask closer to its surface I could see the astonishing detail of the growth before me. Not any two square centimeters were alike. It was a natural, living, variegated tapestry of near infinite variation. I saw countless, rather small barnacles waving their feathery cirri in the ongoing current, embedded in the gray, mottled carpet of living creatures that included anemones of many varieties as well as crinoids, mollusks and various intricate sea fans, just to name a few species.

Were it not for the habitat, none of these organisms would have had the opportunity to thrive. Without the structure, there would have been only the sand below – not an advantageous surface on which to build, grow and pass from generation to generation. It was the presence of the manned habitat that allowed these billions of creatures to live out their lives. What I witnessed was a surface area of more than 660 square feet completely coated with tiny organisms, and there was not a single patch of its surface that was not totally overgrown.

Here was a nearly infinitely complex living layer of both host and symbiont existing together in a kind of furtive tension and meticulous competition.   Each wave and current literally represented the waves and currents of naturally determined chance and the natural decision of who would live and thrive here and who would die, all on a microscopic scale of incredible complexity.

But on the opposite side of the scale, from the microscopic to the macroscopic, the tropical fish also thrived on the living carpet before me.   Although not a main stay of their diet, the grunts, angelfish, parrotfish and tangs also fed off this carpet and were a real part of this living city.   They glided and hovered over its surface like immense blimps and waterships of the aquatic void.

Every future human habitat should have the opportunity to have the same fate undersea. Instead of daily cleanings (except, of course, for the windows), the skin of the habitat offers the opportunity for countless life forms to find a chance to live and thrive. Rather than worrying about the appearance to human eyes, habitat exteriors are best left as nurseries of undersea life and, in so doing, they become endlessly fascinating classrooms that sit as perfect and welcome ecological niches within the blue dominion.

After an hour of cleaning and brushing, Claudia would return inside, shower and change to begin our daily operations.

Aquanauts come to the undersea world to live and work in a wide range of applications. Of my own work, there has been undersea agriculture (which takes place inside the habitat, as opposed to aquaculture which takes place outside the habitat in the aquatic environment) as well as education and outreach activities. Aquanaut duties have also included pharmaceutical studies, research in human physiology, undersea communications and geology. Aquanauts are frequently at work in the petrochemical trades, in space analog research, oceanography, marine zoology and in areas that range widely over most scientific disciplines.

The workday is typically intense, beginning right after breakfast, non-stop until a break for lunch. The aquanaut's day is typically scheduled well in advance and laid out on a daily time grid broken up into half-hour, or even 15 minute, blocks. Most missions are time intensive because every minute keeping a crew undersea is relatively expensive. For obvious reasons, the planners and sponsors want every minute to count toward realizing the objective.

This is not always the case, however. Since the *Jules' Undersea Lodge* was opened in 1986 by pioneer explorer Ian Koblick, many hundreds of aquanauts, and even aquanaut families, have been able to enjoy living in the sea for nothing more than the unique peace and tranquility of the experience.

For the working aquanaut, lunch is a welcome break in a very busy day. Lunch comes in a variety of choices, most of them made weeks before the mission began. As was mentioned previously, all meals were delivered to the habitat in watertight boxes. The food was double-wrapped in plastic bags on the surface and popped into the plastic transport case that had small, pressure valve-knobs screwed into it near the handle. Before entering the water, the duty aquanaut would assure the case was latched down and the valve was tightly closed. A fifty pound lead weight would then be clipped to the case bottom before it was lowered into the water. The weight would counter the air voids in the case and render it slightly negatively buoyant. Without the weight, the aquanaut would never be able to push it underwater!

Mission control always announced that a delivery was underway so the aquanauts in the habitat could await the arrival at the moonpool. As soon as the case arrived, the aquanaut inside grabbed the handle and pulled it onboard. On busy days, the duty aquanaut usually waved hello from underwater and either immediately turned back to the surface or poked their head in and waited for a return package in the case.

As soon as the case was pulled from the water, the first order of business was to turn the valve and equalize the pressure. If that was not done, the case could never be opened at all. The case was sealed on the end of the dock under a sea level pressure of 14.7 pounds per square inch (psi). But when it arrived at the habitat, it was now under a pressure of 23.52 psi. A difference of a mere 8.82 psi does not sound like much, but if you multiply that over the area of the case, it meant that it would require a force of around two and a half tons to open it! To gain access to the case, the aquanaut somehow had to offload the elephant-mass of pressure on its lid. That was accomplished simply by twisting the little vent knob and allowing the habitat air to rush inside, equalizing the pressures. The whole process took less than a minute.

THE DUTY AQUANAUT DELIVERS DINNER

While our meals underwater were planned well in advance, sometimes mission control would take pity on what were sometimes referred to as the 'wet scientists' and would order a meal from a local bistro. What a pleasant surprise it was to open the yellow lid of the underwater case and find a hot pizza or gourmet sandwich inside!

The afternoon began with the same intensity as the morning schedule – no rest or nap time for the aquanaut on a daily mission, I'm afraid. When our missions ran over weekends, sometimes Sunday mornings were allowed as 'personal time'. But typically, underwater mission time was so valuable that there were activities crowded into every underwater day.

Some aquanaut missions require many hours spent outside the habitat, working directly in the ocean environment. Other missions require almost no time outside. I have served on both kinds. But typically, most aquanaut missions require a mixture of work schedules. Preparing the habitat for a mission then reconfiguring it requires almost all the working day to be spent in the water, and the reverse of that during the core of the data gathering phase.

The end of the professional day comes whenever the mission schedule determines that it will come. I have worked missions that demanded a 16-18 hour professional day, allowing for 6-8 hours of sleep with no personal time. But most missions below the surface or above can not task people to that degree for long and remain efficient or even safe. Hence, most aquanaut mission professional days run between 8-12 hours max.

I typically ended a day of concentrated effort with a swim outside, and I always looked forward to that respite to stretch my legs and pump some air through my lungs. At the close of a long day, I would let mission control know that I was about to embark on a swim, known in the parlance as an 'excursion', outside the habitat. I would drop into the moonpool, push away from the habitat and swim hard, stretching my leg muscles and feeling the energy of the air pumping in and out of my lungs. Although I took great pains to design the habitat for maximum livability, in the end, a tight space is a tight space and there is no amount of design intellect that can do away with a small floor plan. Hence, the ability to go outside, flex the fins and run up my heart rate was of paramount importance to me once or twice per day.

Evenings were always looked forward to by everyone with some anticipation. While being an aquanaut crewmember is about as exciting as it gets, the time for a break in the action was always welcome. One crewmember fixed dinner (our typical microwave affair) while the other straightened the habitat after a long day.

It is surprising how quickly so much clutter can accumulate in a small space over a single working day. And after being trained in safety in the US Navy and as a Spaceflight Systems Safety Specialist for so many years, it was second nature to me that clutter and items out of place are also a safety hazard when emergencies always seem to arise without warning. Unfortunately for my fellow crew members and those who had to serve with me as their Mission Commander, I am a neat-freak and a safety fanatic, and will always be.

AQUANAUT
RICHARD B. SCHEALER

Dinnertime focused on the microwave oven that was neatly tucked into a corner near the main power distribution panel. The whole array of controls, venting, oven, cabin conditioning plenums and controls, and even our movie player, was fitted into an exceptionally small space that was tightly and most efficiently designed and constructed by a retired NASA engineer and veteran of nearly all manned spaceflight projects, Richard B. Schealer, and the very talented and enterprising Bill Mandeville. Without their selfless services and amazing talents, the habitat would literally never have left home base!

I have always loved dinnertime in any habitat I have been privileged to occupy. The most fascinating thing about dinnertime is watching the 'sunset' undersea. Of course, there is no orange globe sliding below a horizon. But, much like sunrise, sunset occurs gradually. The blues and greens deepen to a dark emerald carefully layered within rich cobalt, fading gradually to black. It has always occurred to me that on a sunlit day the evening progresses as though the habitat was slowly sinking into the deeper ocean depths. Six o'clock in the evening at 30 feet appears somewhat like 200 feet at high noon.

Sitting at our dinner table while watching the darkness progress undersea was a most restful and appropriate end to a busy day. The twilight environment seemed to underscore the difference between the hurried, always pressing, perpetual urgency of the light-filled world above with the softness, deliberate slowness and the calm eternal balance of the undersea world. Nothing happened here with compelling urgency. Instead, the undersea regions of the earth unfold with a serene and measured tempo.

Like the rare lunchtime treat from the local bistro, on one occasion we were treated to a most extraordinary dinnertime adventure. We had situated our habitat on the seafloor directly between two habitats that had been in place for years – the *Jules' Undersea Lodge* and *MarineLab*. For the first time ever, there were three fully functioning, independent, manned undersea habitats within swimming distance on the seafloor. Even as of this writing, it was the largest undersea community ever assembled.[2]   To celebrate the occasion, Ian Koblick and his crew held a dinner at the *Jules' Undersea Lodge* – just a 30 meter swim from our moonpool. As it was to be a black tie affair, there were those who came attired in paper cut-out black ties attached to their t-shirts for this grand event.

As far as I know, the event was also the first undersea community gathering in history – with the residents of three independently functioning undersea habitats with three dissimilar crews on three independent missions meeting on the ocean floor to celebrate a community wide affair. It was certainly one of the most extraordinary moments of my life.

I will never forget the swim back home that evening after the dining and celebrations had ended. As Claudia and I slowly kicked our way back to our habitat following the beam of our lights, I was struck with the momentousness of this occasion. I was not only blessed to be living and working undersea not just in a solitary habitat on a single mission, but it was in a functioning undersea community – the very first of its kind ever on earth!

---

[2] While Jacques Cousteau launched more undersea structures during Conshelf II, the *Starfish House* and *Deep Cabin* were the only two fully functioning habitats, and all were associated with a single mission.

Retiring in an undersea habitat at the end of a long day is just as spectacular as any moment during the day. We would sometimes sleep three in our habitat because the night view was just so incredible. When word got out, we typically allowed one extra visitor to bunk in with us. These visitors were no strangers, of course. They were always fully qualified staff aquanauts, astronauts, marine architects or other local marine scientists who came to call and give us the added benefit of their knowledge.

My last official act as Mission Commander each day was always one fixed on safety. I required that every crewmember sleep with his light, dive mask and small air bottle by his head and close at hand. I also briefly discussed evacuation procedures and warned everyone to be diligent and hurry, since I would always be the last to leave. At that point, I called up to Mission Control, exchanged our last data set for the day and bade them good night.

Although at this point we turned off our cabin lights, we left some external habitat lights turned on. From the outside, I had seen the habitat with its external lights all on at once. The habitat looked exactly like an alien spaceship under the shimmering surface waters. Its glow was exclusively green and quite bright. It was quite a spectacular sight from the vantage point of the surface.

After the cabin lights were out, we would settle down in the comfy chairs that I had installed specifically for lean-back sleeping comfort and the third aquanaut would spread out on the deck behind the chairs. On the console in front of the right chair were the controls for the lights in front of the station and at the rear. We could switch them to several illumination patterns and one in particular pointed from the side of the habitat just below the windows to a 45 degree arc up. This allowed the light beam to hold just off the plane of the window and slightly forward so as not to induce glare. And in this plane the nightly spectacle began.

When we launched the habitat and began submerging it, we noticed that our plan to make the window bolts air-tight had somehow gone awry. We had subjected the pre-launched habitat to hours of pressure testing and had determined that there were not any significant leaks. Further, upon launching and pressurizing it to pier side depth (20 inches of seawater), there seemed to be no leakage issues that we could detect. But once we submerged the full structure, any minute air leak became immediately apparent. When fully submerged, the habitat took on the appearance of a giant aquarium airstone! Every window bolt passed a thin stream of fine, spinning air bubbles that followed the curvature of the bolt threads![3]

The solution to this was painfully simple. Because of the constant worry about window integrity, I did not allow equipment to be installed near the window frames and all the bolts were required to be left fully exposed. We simply opened a tube of silicone caulk and squirted it onto the bolt shafts. Pressure did the rest of the magic – sucking the silicone along the length of the bolts and near instantly sealing the leaks! But as we worked, we fortuitously decided to leave half a dozen small leaks along the lower side of the front windows. These small, delicate, spinning air bubble streams were quite fun to watch. I reasoned that if they caused some problem in the mission, of which I could not imagine, then we could seal them later. It turned out to be one of the greatest calls of the entire mission!

At night, with the cabin lights off, under the soft green glow of our instruments while we lay in our reclined seats in the darkness of our undersea world, the lights and bubbles began to work their magic. The fine, spinning bubbles were incredibly beautiful to watch in the darkness, all playing together in their individual streams. The Atlantic Silversides showed up and began frolicking in the streams of bubbles, darting in and out as though somehow they could find a morsel to eat. And so it was that within minutes of lights out, the bubbles and fish began their remarkable dance together in the darkness.

But, soon enough, the drama deepened even more. Like a fast moving shadow, a larger fish would dart in from the void and pick off his dinner from among the dancing frenzy of the tiny

---

[3] The 'ambient pressure' habitat was internally pressurized to the depth of the moonpool. Therefore, the air pressure inside the habitat was slightly greater than the outside, with a max relative positive inside pressure of 3.1 psi. Therefore, the habitat could never allow water to 'leak in' like a submarine that is pressurized at 1 ATM. If an ambient pressure habitat springs a leak, it is by definition a leakage of air outside, never a leakage of water inside.

Silversides. At this act of natural predation, all the smaller fish would vanish in a single instant, leaving the bubbles alone streaming toward the surface. But within moments, the first of the smaller fish would be attracted back to the eternal rising streams of glassy bubbles and the drama would slowly begin again.

While each act had the same dramatic conclusion, the individual performances were never the same, which is what kept us glued to the massive, transparent screen night after night. On some occasions, there was but a single predator. On others, there were many. Some nights the tiny fish never showed up at all, leaving the larger fish to cruise the lights with their own slow moving dance in and out of the bubbles.

Eventually, even the best movie fails to hold the eyelids open. As soon as I saw that every eye was closed, I would switch off the outside lights, pull up my blankets and try to go to sleep. Once sleep was attained, I found it to be deep and manifestly restful. However, turning my mind off was never easy. I always felt responsible for every life entrusted to me for any particular mission. I always worried about each crew member and even for the quality of the mission and its impact on the ones who were paying for it. My mind turned loose of these responsibilities slowly, almost reluctantly. A type 'A' personality and an underwater mission are two powers inimical to rest. But rest finally came – it always came. Even against my best efforts, merciful sleep ended each day with the finest rest on earth.

It seems to me that my last words were always a breath of prayer – prayer for safety, for greater wisdom and most of all, they were prayers of profound thanksgiving for the most incredible opportunity of anyone's lifetime. But the most awesome thing of all was the knowledge that the very next morning, I could awake and do it all again!

A BRIEF HISTORY OF MANNED UNDERSEA EXPLORATION

"Great things are done when men and mountains meet."
William Blake

It is sheer speculation when we assume what the first civilized humans must have thought when they looked at the vast expanses of the oceans of the world. It is certain that they considered them dangerous. It must have been quite rare to find someone who lived around aquatic bodies who did not know of a human drowning or tragedy connected to the water. Even calm bodies of water leant the possibility of death any time humans dared to set foot into their void. So we presume that the first humans thought of the great deep with fear. It is thus not surprising that when we look at ancient manuscripts that they would reflect terror, uncertainty and even chaos at the mention of the deep.

In ancient literature, such as in the Judeo-Christian Bible, the underwater world is frequently described in dread as 'the deep'. Indeed, it is recounted in almost mysterious terms. It is a place representing lostness and the most extreme form of isolation. King David referenced it twice:

"Deliver me from the deep and do not let me sink; may I be delivered from my foes and from the deep waters. May the flood of water not overflow me nor the deep swallow me up..." Psalm 69:14-15

"If I take the wings of the dawn, if I dwell in the remotest part of the sea, even there Your hand will lead me, and Your right hand will lay hold of me. If I say, 'Surely the darkness will overwhelm me, and the light around me will be night'..." Psalm 139:9-11

The prophet Ezekiel also wrote, "For thus says the Lord GOD, "When I make you a desolate city, like the cities which are not inhabited, when I bring up the deep over you and the great waters cover you, then I will bring you down with those who go down to the pit, to the people of old..." (Ezekiel 26:19-20) And the first submariner, the prophet Jonah, recounted from his experience, "Water encompassed me to the point of death. The great deep engulfed me; weeds were wrapped around my head." (Jonah 2:5)

And then there is my favorite Biblical reference: to the Apostle Paul as history's first aquanaut when he states in 2 Corinthians 11:25, "Three times I was beaten with rods, once I was stoned, three times I was shipwrecked, **a night and a day I have spent in the deep.**"

But it was inevitable that some human would look at the transparent deep of the undersea regions and not consider it a threat or necessarily dangerous. Some humans would look at the invariant beauty of the undersea world from the surface vantage point and, just because of its overwhelming allure, want to enter there. But the reasons for that impulse were not just aesthetics. There were other motives to want to submerge beneath the great waters of the world, from economic to military.

Predating even Judeo-Christian literature are indications that right at the dawn of civilization, around 4500 BC, ocean side communities such as Mesopotamia, China and Greece were involved in free diving activities for food gathering. Around 1000 BC, perhaps the earliest mention to a diving apparatus is an Assyrian fresco that depicts men swimming underwater, using a kind of breathing mechanism.

In the Iliad, Homer describes the use of divers in the Trojan Wars for military operations. And around 1000 BC Homer also mentions Greek sponge fishermen who dove down to 100 feet by holding onto rocks. Half a millennium later a Greek diver by the name of Scyllias and his daughter Cyana used reeds as snorkels and cut the mooring lines of Persian King Xerxes' ships.

THUCYDIDES

Notably, in 414 BC, historian Thucydides tells of the use of Greek military divers sent to clear obstacles from the harbor of Syracuse. 54 years later Aristotle describes an air supplied diving bell for sponge collection in his *Problemata* (not a name I would want to apply to any work during an expedition). He noted thus: "...in order that these fishers of sponges may be supplied with a facility of respiration, a kettle is let down to them, not filled with water, but with air, which constantly assists the submerged man; it is forcibly kept upright in its descent, in order that it may be sent down at an equal level all around, to prevent the air from escaping and the water from entering...."

It is reported that Alexander the Great was so fascinated by the undersea world, that in 332 BC he submerged in a diving bell to personally observe his military divers removing underwater obstacles during his siege of Tyre, now modern day Lebanon.

A CATALOGUE OF EARLY UNDERSEA
EXPLORATION EQUIPMENT

The first laws governing undersea work were set in place in 100 BC around major shipping ports in the eastern Mediterranean to cover wage scales for divers that acknowledged that hazard increased with depth. In 77 AD Plinius the Elder reports the use of air hoses by divers. In 1300 AD Persian divers were using eye goggles made from polished shells.

The first true technological advance came in 1500 when Leonardo da Vinci designed the first known SCUBA apparatus. The drawing appears in his famous *Codex Atlanticus*. His design was so ingenious that it actually incorporated an air supply and a buoyancy control device. Later, it is thought that he abandoned this idea in favor of the less technically challenging diving bell. There is no evidence that Leonardo ever progressed beyond his drawing.

In 1535, Guglielmo de Loreno developed what we consider today to be the first true diving bell. 1650 saw Von Guericke invent the first effective air pump and in 1691 Edmund Halley patented his diving bell connected to barrels of submerged air.

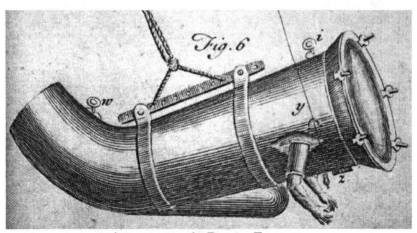

LETHBRIDGE'S DIVING ENGINE

In 1715, inventor John Lethbridge had to find a way to feed his 14 children. So he invented what he called his 'Diving Engine' – an oak cylinder that was supplied with air from the surface. It was, of course, the first diving suit. His arms extended through holes in the side of the suit sealed by greased leather cuffs. What did John want to do each and every time he returned from taking his diving engine under 12 fathoms for up to half an hour at a time? Don't ask his wife.

It was right about this moment in history that machines and machine building techniques began to advance rapidly. And with that advance, the ability to create new devices began to emerge, hosting a wide range of novel approaches to many aspects of exploration, particularly undersea.

In 1776, the first authentic attack by a military submarine occurred in New York Harbor. The seven foot tall wooden

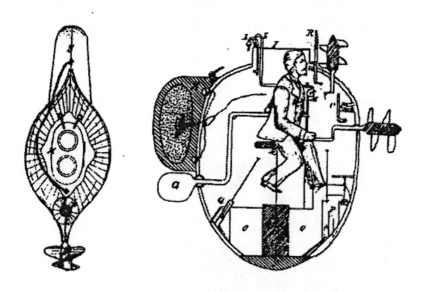

THE TURTLE

submarine *Turtle* was shaped like an egg with a windowed copper conning tower and had two hand cranked propellers. On September 7, captained by Army volunteer Sergeant Ezra Lee, *Turtle* attacked *HMS Eagle* which was moored off what is today Liberty Island, but it could not manage to bore through the *Eagle's* hull and deposit the 100 pound time bomb it was carrying. When he attempted another spot in the hull, Lee bounced off the ship and bobbed to the surface, eventually abandoning the endeavor and fleeing to safety.

In 1823, Charles Anthony Deane patented a helmet used to protect firefighters from smoke. Its secondary purpose was to supply divers with air while working underwater. Five years later,

Dean and his brother marketed the same helmet with a diving suit for exclusive use underwater. Two years following that, another inventor, Augustus Siebe, mated the helmet to a watertight arrangement that included a rubber suit. In 1843, after exclusive use of the new diving suit for military purposes, the Royal Navy established a diving school to train military divers. Then, in 1876, Henry Fleuss developed a dive apparatus that was self contained and used compressed oxygen.

Between 1878 and 1912 three key books were published that contained information on how pressure affected human physiology: Paul Bert's *La Pression Barometrique*; John S. Haldane, Arthur E. Boycott and Guybon's *The Prevention of Compressed Air Illness*; and the *US Navy Dive Tables* which were developed from the earlier publications and studies. In 1917, the US Navy also introduced the Mark V Dive Helmet used throughout World War II for all Navy undersea diving work.

BEEBE'S BATHYSPHERE

In 1930 William Beebe descended 1,426 feet below the surface of the ocean in a bathysphere. In the same decade the modern dive mask and fins were invented; the mask by Guy Gilpatric and fins by Louis de Corlieu in 1933. Beebe and Otis Barton also made another deep water excursion in his bathysphere to 3,028 feet.

In the early 40's Italian divers routinely used closed circuit diving equipment to place underwater explosives beneath British naval and merchant ships. But the most important apparatus ever devised to enable humans to live and work in the sea was invented in 1942-43 by Jacques Cousteau and Emile Gagnan. They introduced the 'Aqua Lung', an open circuit device that allowed divers to breathe from a bottle of compressed air worn on their back. Cousteau and Gagnan's invention of the Self Contained Underwater Breathing Apparatus (SCUBA) literally opened the doors of the undersea kingdom to the average human. From 1943 to date, the undersea region has been available to nearly every human with an interest to explore the submerged world.

COUSTEAU AND GAGNAN

The first wave of SCUBA explorers entered the sea by the thousands after the Aqua Lung was marketed in France (1946), then Great Britain (1950), Canada (1951) and the United States (1952). Gagnan faded out of public sight almost immediately. But Cousteau parlayed his invention into international fame and became the best known and undisputed spokesman for all of manned undersea exploration.

With the legions of undersea explorers also came calamity. The use and misuse of SCUBA equipment resulted in the formation of the YMCA SCUBA certification courses in 1959 followed by the National Association of Underwater Instructors (NAUI) in 1960 and the Professional Association of Diving Instructors (PADI) in 1966.

In 1960, the submariner achieved a monumentally important objective when Jacques Picard and Don Walsh descended 35,820 feet to one of the deepest points in the ocean in the bathyscaphe *Trieste*.

In 1961, President John F. Kennedy ignited the imagination and passion of the entire world when he challenged his countrymen to travel safely from the earth to the moon and back before the end of that decade. His challenge reverberated throughout the exploration community and the undersea region was also immediately caught up in the rush to explore the frontiers. 1962 was an extraordinary year for human exploration of every frontier. Man's settlement of the undersea regions of his planet began that year on the French Riviera, and here the history of the aquanaut began.

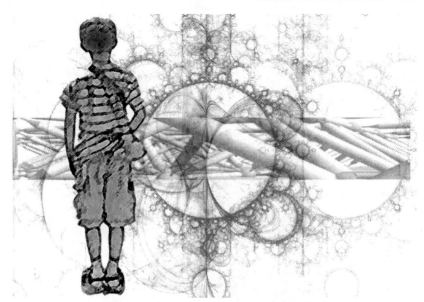

A HISTORY OF AQUANAUTS LIVING AND WORKING IN THE SEA

"It is our destiny to move out, to accept the challenge to dare the unknown..." **Louis L'Amour**

I n any fair historic assessment, US Navy Captain Dr. George Bond is considered the Father of the Aquanaut. The idea that man could live and work undersea for long periods of time – days, weeks or months – was originally a military initiative. Specifically, it was Captain Bond's idea. The deep irony of Captain Bond's certain paternity lies in the distinct possibility that his idea of *how* to make this happen may later be demonstrated to be the first misdirected start in permanent human dominion of Aquatica. That concept will be explored in depth in a later chapter.

George Bond was a physician practicing in the small community of Bat Cave, North Carolina. He became concerned about a possible shortage in animal protein to the world's rapidly increasing population. Dr. Bond blended his interest in the sea and reasoned that perhaps the vast planetary oceans could provide a solution to the much feared and impending protein deficit. He joined the United States Navy in 1953 and his career led him to advance to the Assistant Officer in Charge of the Naval Submarine Medical Research Laboratory at New London, Connecticut, in 1957.[a]

And thus it was that Dr. Bond joined the US Naval diving community. At that time, the Navy had obvious interests in undersea involvements considering its military mission. Its greatest undersea resources were invested in submarines – moving platforms that projected the nation's military might and geo-political influence all over the planet. But the Navy also had an interest in undersea salvage, underwater reconnaissance and submerged, unmanned listening arrays. Bond's official position was an interest in significantly extending the Navy's role by establishing permanent undersea bases. But his true intent actually extended far beyond that relatively limited beginning.

The Navy had accumulated a wealth of knowledge in deep diving and decompression. Key to this process was timed intervals at deep pressures, followed by long periods of decompression, which is defined as a gradual lowering of breathing gas pressures so that dissolved gases in the blood do not evolve out into the blood stream as bubbles, causing internal injury. These decompressions were usually in metal chambers aboard ship at the end of a dive. Experiments had been carried out using not only oxygen and nitrogen mixtures such as were found in the air, but also other gasses in different mixtures to solve a myriad of medical problems, including oxygen and nitrogen toxicity under pressure. But it was soon discovered that repeated decompressions had their own long term and inherit risks.

Yet Bond reasoned that if a diver could be taken down to pressure in a habitat, he could live and work out of that habitat for days, weeks or months without having to surface until the end of

GEORGE F. BOND

U.S. Navy Captain George F. Bond, MD. can be called the father of the aquanaut. He set out with a specific purpose, according to his 'Project Genesis' venture. Bond stated that enabling man to live and work in the sea was an important step in attaining man's dominion of the sea as he noted was promised in the Biblical account of the Genesis. His efforts in saturation diving seeded all the subsequent first steps of humans living and working in the sea from Edwin Link's Man in the Sea projects, which led to the first aquanaut, to Cousteau's CONSHELF and the US Navy SEALAB projects that he helped manage.

the project and thus not have to endure the dangers, risks and discomforts of repeated compression cycles. This was a process called 'saturation diving', in that the diver 'saturated' his body with gasses under pressure to the working depth. In this sense, there would have to be only a single decompression no matter how long the diver stayed down.

But in reality, Bond's interests extended far beyond that. He felt that if humans were to survive on the planet, they would have to enter the underwater world and remain there to explore, observe and harvest the wealth of the oceans.[b] He called his program 'Project Genesis' and boldly declared its lofty vision as "… expansion of man's ability to utilize the products of the marine biosphere which make up nearly three quarters of our Earth."[c]

Bond named his project Genesis because he felt it was an important step in mankind's attaining 'dominion of the sea' as promised in the Biblical book of Genesis. In his own words, Bond was seeking the permanent presence of humankind in the ocean and he was using the US Navy as his vehicle to make that happen. This, by any means, was not a fault. Indeed, he saw it as an opportunity and capitalized on it without delay.

Bond carried out Project Genesis in five phases, beginning in 1957 with rats under saturation pressure and progressing to human subjects, all in shore-based hyperbaric chambers. Six years later, the last experiment was complete and, as a result, Bond had devised a schedule of gas mixtures and decompression schedules at saturation that opened up the undersea world to saturation diving at depths that Bond himself described as limited to 'any point on the submerged continental shelf".[d]

It is the purpose of this book to describe Dr. Bond's work as historic, groundbreaking, landmark research. And it was. Unfortunately, it also led the ultimate human settlement of the undersea world down a misdirected path of exploration. In the end, the central idea set forth in Project Genesis set back the permanent human settlement of the seas for decades. While Bond was headed in the right direction, the end of the pathway was

JACQUES YVES-COUSTEAU

Jacques Yves-Cousteau opened the undersea world to every human in 1943 with his invention of the Self Contained Underwater Breathing Apparatus (SCUBA) along with co-inventor Emile Gagnan.   He began the quest to carve the way for humans to live undersea with the brilliant CONSHELF Expeditions in 1963 – 1965. He then went on to explore, film and write about the undersea world so that, regardless of whether a diver or not, everyone on the planet shared in the great quest through his eyes.

profoundly limiting. That opinion is discussed in detail in a later chapter.

Humans had ranged underwater for long periods of time since the advent of the first military submarines. With the invention of the first nuclear submarine by Admiral Hyman Rickover, the USS *Nautilus* in 1954, the engineering capacity to remain submerged actually exceeded the logistical capacity to do so. The on-board nuclear power plant supplied so much energy that oxygen for breathing could be split from seawater and carbon dioxide could be scrubbed from the air almost indefinitely.

But, in 1962, the line was drawn between who would be considered a submariner and who would be designated as an aquanaut. A submariner was defined as one who roamed the seas in a moving platform. An aquanaut was one who lived and worked undersea in a fixed habitat structure for more than 24 hours without surfacing. They were no more alike than a truck driver and a farmer.

The first human aquanaut was Robert Stenuit who lived onboard a tiny, one man cylinder-habitat for 24 hours during a saturation and decompression experiment in which he was participating. For the first time in human history, man submerged into the oceans to settle down and do work – not to race from one point to another; not to do damage to an enemy; not to enter for a few hours and retreat back to the surface – but to enter specifically to stay for protracted periods of time. It was not yet time for permanence, but it was time to start thinking seriously in that direction.[e]

Robert Stenuit, who had achieved the designation of 'First Aquanaut', was marching off smartly to his cylinder on the shoulders of several key giants, including George Bond. Stenuit was following Bond down the path that he had paved – saturation exposure – which had actually allowed the science of Aquanautics to begin years before the historic dive. Stenuit was connected to Bond by American inventor and explorer Edwin A. Link.

Link was an undersea archeologist who dreamed of keeping crews for his peaceful undersea exploits on the seafloor for extended periods without the need to surface and decompress. He seized on Bond's data and, with help from the National

IAN G. KOBLICK

While Jacques Yves-Cousteau opened the oceans to mankind though his invention of SCUBA, Ian Koblick opened the undersea world to aquanauts. In a brilliant strategy and bold business plan, Koblick unlocked the frontier with his *Jules' Undersea Lodge* and the adjacent undersea laboratory, *MarineLab*. Koblick's genius allows everyone who wishes access to living and working in the sea. Through his single-handed and determined efforts, more humans have been granted access to living and working undersea as aquanauts than through any other single endeavor. Ian was also the habitat designer and program manager for the important *La Chalupa* Missions (1972-1975). He has steadfastly maintained these open doors by his own unrelenting initiative for more than two decades, and they remain the world's only freely available portals to Aquatica. Along with James Miller, Koblick is the coauthor of the essential and peerless work, *Living and Working in the Sea.*

Geographic Society and the Smithsonian Institution, devised a human rated aluminum cylinder designed to be lowered off the side of a boat to depths of up to 400 feet. The tiny cylinder measured 3 feet in diameter by 11 feet in length. It was so small in fact, that is was positively claustrophobic and its occupant could only slide in and lay reclined to wait out the event.

Link called it: Man In The Sea I, or MITS. Link himself became the cylinder's first human occupant in August of 1962 at Villefranche Bay on the French Riviera. He made two dives in the cylinder: one to 35 feet for 2 hours and another to 60 feet for eight hours breathing mixed gasses. But Link's tests were only the warm-up leading to the first certified aquanaut mission in history.

On September 6, 1962, Robert Stenuit entered the MITS cylinder and was lowered to 200 feet. The plan was for him to remain in the cylinder for 48 hours. Once he reached depth, he pressurized the chamber to ambient pressure, opened the hatch and exited for a series of experimental dives. But at 24 hours and 15 minutes, the dive was aborted for technical reasons. On the way to the surface, Stenuit began decompression in the cylinder even before it was hoisted onboard the boat. He also faced some pain in his wrist which forced his ultimate stay in the cramped cylinder for 65.5 hours.

When Stenuit crawled out of the cylinder, he exited the world's first aquanaut and did so specifically because the work of George Bond had paved the way.

The Great Race Begins

Bond's milestone work had an almost magnetic quality to it. Not only had Link absorbed every word and detail, but so had French explorer and inventor Jacques Cousteau. At the same moment the National Geographic Society and the Smithsonian Institution were working to support Link, the French government's Office Francais de Recherches Sous-Marines (French Office of Underwater Research), under the direction of Jacques Cousteau and team, were designing their own undersea habitation program called Conshelf I and II.[f]

JAMES W. MILLER

A true champion and pioneer of undersea habitation, Dr. James Miller is the founder of the brilliant Tektite Missions (1969-1970) which were his brainchild. He served Tektite I as Deputy Project Director and Tektite II as Project Director. He also served the SEALAB II missions as a civilian Director of Psychological Programs. Dr. Miller was the Director of Ocean Technology for the Department of the Interior prior to the formation of NOAA. He was also an aquanaut crewmember for the *La Chalupa* Missions in 1973-1974. Dr. Miller is the co-author of the most important book written on the first phase of human undersea habitation, *Living and Working in the Sea*.

The term 'Conshelf' referred to the vast areas of moderately deep waters that surround the world's continents, called the continental shelves. These areas represent an astonishing amount of territory unclaimed by permanent human activities. For example, the continental shelves of the continental United States and Alaska consist of nearly 1.5 million square miles of seafloor, just less than half the territory of the United States. Cousteau's bold vision was to engineer access to these areas by building the equipment and learning the techniques to enable the process to begin.

His first habitat in the Conshelf I program was called *Diogenes* – an 8' by 17' cylinder held to the ocean floor by chains and iron.[4] It touched down on the seafloor at 32 feet off the coast of Marseilles, France, on September 14, 1962, just a week after Robert Stenuit had become the first aquanaut in history. But what a difference just a single week made! Instead of one man cramped into a tiny tube, the *Diogenes* was, by comparison, a four star suite! It offered history's second and third aquanauts, Albert Falco and Claude Wesly, a luxurious pair of bunks and plenty of leg room, plus storage and eating areas. Instead of a single day visit, they lived and worked out of their habitat for a week.

Granted, Stenuit's depth was much greater – 200 feet compared to a modest 32 feet. But the remarkable story, in review of the history, is not at all depth or relative luxury, but the intensity of the international race that had been sparked by George Bond when he effectively handed over the keys of living and working in the sea. International records and firsts – as were ongoing in space exploration at the exact same moment – were only separated by a matter of days.

After the clunky, lumbering and crudely appearing *Diogenes* was winched back to the surface in late September 1962, Cousteau launched into a furious effort to construct his ultimate human penetration of the sea, which he called Conshelf II. The habitats and machines of Conshelf II were light years ahead of the *Diogenes* in both form and function. In a mere nine months, Cousteau's machinery shops and technical team had constructed

---

[4] The habitat was aptly named after the historic character by the same name who lived in a tiny barrel on the streets of Corinth.

MORGAN WELLS

In addition to designing and operating his own unique undersea habitat, the *Baylab*, Morgan Wells has more experience as an aquanaut in different habitats than any other individual. He has served as an aquanaut on SEALAB (1 mission), Tektite (2 missions), Edalhab (2 missions), Hydrolab (more than 12 missions), La Chalupa (3 missions) and USIC (2 missions). Morgan Wells has served in the capacity of experimental diver, aquanaut, operations director, systems certification authority and all phases of design reviews.

structures that amounted to the first temporary 'colony' on the ocean floor.

COUSTEAU'S STARFISH HOUSE

For the purposes of this work, the term 'colony' references specifically a permanent human presence. Cousteau's 'colony' was anything but permanent. It was not designed to be permanent and it did not remain on the ocean floor but for a single month, then the aquanauts returned to base ashore, their duties over and the experiment completed. Hence, Conshelf II represented the first test of multiple habitats and undersea structures working together

CHRISTOPHER OLSTAD

Christopher Olstad is the only professional aquanaut that has reported to work regularly for the past 25 years as a working aquanaut. He has accumulated more hours living and working undersea as a professional aquanaut than any other human being. Chris is the manager of the *MarineLab* Undersea Laboratory at MRDF in Key Largo, Florida, which has logged more hours than any undersea laboratory in history. Through Chris' management, *MarineLab* is also the world's longest lived manned undersea habitat, in continuous use since 1985. Chris is as close as any human has ever been to a true Aquatican.

as a unit more than an actual human settlement. It is essential not to underestimate the importance of Conshelf II, but equally essential not to classify it as something it was not – and it most certainly was not anything even resembling a permanent undersea colony.

Nonetheless, Cousteau succeeded brilliantly in both form and functional design unequaled by any other undersea mission - even to the day of this writing. His Conshelf II habitats and structures were strikingly beautiful and seemed to fit perfectly in their surroundings. It was as though an undersea architect visited Cousteau's drawing boards and gave them a sweeping visual sense of the future.

There were three main undersea structures associated with Conshelf II which gave it the appearance of a colony: the *Starfish House* habitat; the *Deep Cabin*; and the *Dive Saucer Hangar*. In the decades since the first aquanaut mission, Conshelf II remains to this day the most comprehensive, well planned and executed, elegant and broadly sweeping human assault as a full fledged practice mission for settling the undersea frontier. Regrettably, it only lasted 30 days.

Conshelf II began its mission on June 15, 1963, in the tepid Red Sea about 25 miles northeast of Port Sudan. Cousteau could not have selected a more hospitable undersea location. The surface water temperature hovered in the mid-80's and visibility was over 100 feet. The carefully selected bottom location included a flat, narrow plain at 36 feet where the large 34 feet in diameter *Starfish House* was staged. Nearby, the 7 foot in diameter and three level *Deep House* was staged at 90 feet over a rather narrow, precipitous ledge. Near the *Starfish House*, a hangar was set to act as an undersea garage of sorts for the complex and relatively high-tech two man *Dive Saucer* submersible.

Yet, all was not perfect in Cousteau's Red Sea community. While the five man *Starfish House* was roomy and air conditioned, the *Deep House* was neither. There, the aquanauts suffered from the humidity hovering at 100% and reported sleeplessness and a loss of appetite. The two *Deep House* aquanauts lived and worked there for a period of seven days, eventually decompressing on mixed gasses and retreating to the *Starfish House* for more decompression.

After four weeks, the brilliant undersea experiment concluded with no injuries and many lessons learned. It was such an exceptional success that 52 year old Cousteau surged onward with an almost inhuman energy for the next phase in seizing the undersea world: Conshelf III. But for Jacques-Yves Cousteau, it was to prove a turning point in his life and in his view of man's permanence in the sea.

While Cousteau had been focused on relatively shallow water placements, American Edwin Link was looking deeper, altogether fixed on the data collected by George Bond. By 1964 he was ready to move on to his Man-In-The-Sea II program.

For MITS-II, Link set his goal at a depth of 400 feet. But Link decided that, based on past lessons, serious undersea habitation programs needed a fast deploying, low transport volume habitat that could be brought on site quickly from anywhere. So he designed and built the world's first, and perhaps only, inflatable habitat called the SPID – Submerged Portable Inflatable Dwelling. It had to provide the aquanauts with warmth and shelter as well as easy entrance and exit capability.

Link's SPID was deployed and tested on the seafloor off Key West for over a month in early 1964. Then, on June 30, 1964, the SPID was lowered to 430 feet in the Bahamas. The first aquanaut, Robert Stenuit, joined by Jon Lindbergh, son of famed aviator Charles Lindbergh, deployed to the SPID for 49 hours, breathing a 3-6-91 oxygen-nitrogen-helium mix. After their mission, their decompression required nearly twice the time that they had lived on the bottom in their inflatable undersea bungalow. Link had successfully proven George Bond's original concept, pushing his divers more than four times Cousteau's limit in a remarkably innovative undersea system.

The United States Navy had entered the manned undersea exploration months before, unannounced, as the wheels of the massive bureaucracy churned out its program that consisted of the Navy's best dive teams and systems engineers. They called it SEALAB I.

SEALAB I was certainly designed by engineers. It looked like a submarine with legs affixed to a pair of pontoons. It was an example of military functionality with little imagination and no attention whatsoever to form or design. It made Cousteau's *Starfish House* look like a hands down winner in undersea architecture.

The Navy's first habitat program got off to a rocky start when it flooded and sank, unmanned, during sea trials in May of 1964. SEALAB I was hurriedly re-outfitted and finally successfully launched in 193 feet of water 26 miles off Bermuda. On July 20, the four man crew arrived on the seafloor and the 11 day mission began. The crew breathed a 4-17-79 oxygen-nitrogen-helium mix – all personally supervised by Dr. George Bond, referred to by the crew as 'Pappa Topside'.

If the Navy had missed the mark in habitat design form, it missed absolutely nothing medically. Under Bond's able supervision, more data was collected on the human capacity to function in this environment than in all previous undersea missions combined. Perhaps unnoticed at the time, Bond discovered the first chink in the armor of his plan, related here by Ian Koblick and James Miller in their one-of-a-kind work, *Living and Working in the Sea*:

"Two important clinical observations were noted during the exposure. There was an apparent slowing of all gross physiological and motor functions, as well as an ability to acclimatize to the body's caloric loss in the gaseous and water environments. It was also noted than when oxygen levels were at 4% or greater, the aquanauts reported an improved sense of well being, a result attributed to the increase density of the breathing gas, which was about 1.6 times greater than sea-level air, causing an increased pulmonary ventilation and a need for increased molecular concentration of oxygen. Within the first 24 hours of exposure at 193 feet, after denitrogenation in the oxygen and helium atmosphere, exposure to compressed air resulted in an immediate and dangerous level of nitrogen narcosis, equivalent to that experienced breathing air at 350 feet. The authors concluded that once the body is essentially denitrogenated, susceptibility to nitrogen narcosis is significantly increased."[8]

SEALAB I's very successful run was terminated early because of a threatening tropical disturbance. But the Navy's efforts leant several valuable insights on living and working in the sea, not the least of which was that mixed gasses under pressure resulted in several major physiological problems. The process seemed to work differently in the remote and extreme environment than it did in the safe, highly controlled confines of the laboratory.

The helium also sapped heat from the human body and made speech all but unintelligible. George Bond's idea of successfully saturating the body with alien atmospheres did not appear to be as straightforward as he would have hoped. Further, the surface support vessels, multiple decompression chambers and weather made the system cumbersome and dangerous. All of the problems seemed to be linked directly to the pioneering nature of the first try, but also to the nature of the unique art and science of saturation diving.

SEALAB II was approved almost matter-of factly, directly on the heels of SEALAB I's success. In an astonishing declaration by 21st century standards, in January 1965 the Navy stated that it wanted aquanauts on the seafloor in a new habitat in a mere eight months! Starting from that declaration, the Navy built and tested all the hardware for SEALAB II as well as selected and trained the crew. And instead of a warm tropical environment, the Navy opted for the relatively deep seafloor at 205 feet in the cold Pacific waters off the Scripps Oceanographic Institution near La Jolla, California. [5]

SEALAB II was designed slightly larger than SEALAB I but with the same absolute disregard for any architectural form. If SEALAB I looked something like a submarine with legs welded on

---

[5] Such a feat in the 21st century would be bureaucratically and legally impossible. From inception of an idea to legally letting the contract requires about 18 months, and that is before the first bolt is turned. Said one frustrated government administrator of the current state of affairs, "Money is no longer a limiting factor. I could have infinite funds and still couldn't get there from here!"

pontoons. SEALAB II looked exactly like a railroad car. The habitat's commissioning photograph looks like bunting was placed around an average, run-of-the-mill tank-car at the railroad yard. In typical military style, function not only displaced form, it completely eradicated it.

But if there were any other disappointments, they were not found inside or in the mission design. Compared to its predecessor, SEALAB II was comparatively luxurious. The habitat sported 11 view ports, heating cables in the deck, a lavatory, an onboard water heater, sinks, toilet, showers, refrigerator/freezer and an air conditioning system. Further, it came with its own 110 foot surface support vessel. And the Navy called on one of its own heroes, Astronaut Scott Carpenter, to lead their undersea program. To further emphasize that Poseidon was truly smiling on the venture, the habitat arrived on the seafloor right on schedule in August 1965.

From Robert Stenuit's tiny, cramped metal tube less than three years earlier, humankind had moved up to SEALAB II with an astounding technological pace. It was the best of times for government sponsored exploration. While SEALAB I and II were under development, the international space race was engaged in earnest and no one knew who would be the first to the moon. It was an extraordinary time when huge government bureaucracies could actually declare cutting edge technical goals, set deadlines in terms of months and actually meet them while not suffering crippling overruns in schedule or cost. Were the Navy to declare the same goals today as they did in SEALAB II, it is no secret what the comparative cost or timeline would be, not to mention the crippling bureaucratic overhead and burden to the project.

ASTRONAUT/AQUANAUT SCOTT CARPENTER IN SEALAB II

With the advent of the US Navy's SEALAB program and the adding of national champion Scott Carpenter to lead the undersea team, it became obvious to even the most casual observer that there was not only an international race to the moon, but a race to occupy the frontier of the oceans as well. Some US Naval Officers involved with SEALAB positively bristle at that suggestion and will say nearly unkind things to anyone who even suggests such falderal.

"Nothing could be further from the truth! There was no race or anything of the kind. We were just doing our duty for God and country!" said one officer at that suggestion in my 1986 article, *SEALAB: Unfinished Legacy* published in the *US Naval Proceedings*.

But the Captain was dead wrong. By the time SEALAB touched down on the seafloor, the race was on and the United States Navy was right in the thick of it – whether they cared to admit it or not.

SEALAB II's ten man crew arrived on the seafloor on August 23, 1965, led by Scott Carpenter. On August 24, Carpenter made history by becoming the first ever Astronaut-Aquanaut. During his initial day on the seafloor, another historic record was set when he spoke with Astronaut Gordon Cooper orbiting the earth in his Gemini capsule: the first seafloor to orbit communication.

But the work load was crushing, comprising some 20 hours of work per day with irregular four hour sleep periods. Other problems cropped up as well – such as headaches and chronic ear infections – many of them similar to the experiences of SEALAB I.

On Day 16 of the mission, only Carpenter remained behind in the habitat while Team 2 arrived. Then on September 26th, Team 3 assumed the mission duties. On October 1st another historical event occurred when the SEALAB II team communicated with Cousteau's Conshelf III crew undersea in the Mediterranean. Then, on October 10th, the highly successful SEALAB II mission was concluded.

But again, the devil was in the details of actually having to live and work in that environment. Here is a review of the results from Miller and Koblick:

"Life in the SEALAB II was highly stressful due to lack of sleep, high relative humidity and communication difficulties caused by the helium environment. Headaches, apparently caused by atmospheric contaminants, and bacterial ear infection, partially resulting from the high humidity and buildup of perspiration under the hoods worn by the divers, were the most frequent medical problems."[h]

Yet SEALAB II was fantastically successful in proving man's capacity to live undersea even under these extreme conditions. That single mission allowed 28 aquanauts to live for a total of 450 man-days, as well as complete 400 man-hours of work, at a 205 feet depth in cold and often low visibility surroundings.

On the other side of the planet, Cousteau had not been idle. In the intervening three years since the end of Conshelf II, Cousteau had poured his resources and creativity into a bold assault on deep waters he called Conshelf III. Just after midnight on September 22, 1965, six aquanauts touched down on the ocean floor off Monaco at a depth of 328 feet to begin a 21 day 17 hour experiment at a record depth. While there, the aquanauts routinely worked at a depth of 370 feet.

True to his past creativeness, Cousteau's Conshelf III habitat was a beautifully designed two story domed structure that was brilliantly engineered for both grace and functionality. The dome itself was 18 feet in diameter with the lower floor committed to diving, sleeping and sanitation while the upper floor was for dining, communications and sported a data collection facility. Cousteau had fixed the dome to a barge that held nearly eighty tons of ballast, gasses and two decompression chamber/lifeboats that could be released to the surface in an emergency. All together, the Conshelf III system weighed in at 140 tons.

Earlier experiments had aptly demonstrated that low density exotic breathing gasses sapped heat from an aquanaut's body and the atmosphere selected for Conshelf III was 2.5 % oxygen and 97.5 % helium. With the outside water temperatures ranging from 50 to 55 degrees, typical for that depth at that latitude, the internal habitat temperature had to be elevated to 90 degrees F.

Uniquely, the Conshelf III underwater breathing system consisted of a fantastically large re-breather. The aquanauts used a hookah system comprised of a pair of hoses attached to the habitat. They inhaled from one hose and exhaled into the other which took the gasses back to the habitat where the CO_2 was scrubbed out and oxygen was added in the pre-described amounts.

Meanwhile, inside the habitat, Cousteau had designed a new cutting edge life support technology where a cryo-generator machine circulated their breathing gasses, precipitated carbon dioxide by freezing the gas out of the air stream (as well as other less desirable gasses) and acted as the dehumidifier and community freezer as well!

But the now well-known problems of living in an atmosphere of high pressure helium began to crop up all over again. While not causing any physiological problems, the annoying gas ruined electronics, leaked at prodigious rates into and out of every human built device and caused human speech to become fundamentally useless for communication. The aquanauts were forced to resort to handwritten notes and hand signals as a primary form of communication.

The Conshelf III experiment concluded successfully on October 13, 1965. Counting decompression time, the crew had spent more than 30 days under pressure.

In 1966, the academic community decided to test the waters. Florida Atlantic University and Perry Submarine Builders teamed up to design, build and launch the *Hydrolab* I habitat. It was placed in the beautiful waters off West Palm Beach, Florida at a depth of 50 feet and was moored to the bottom with chains – a decision that caused seasickness due to the effects of undersea swells. It was later redesigned to sit on the bottom with legs which eliminated the motion.

Hydrolab enjoyed many years of operation and was moved to various locations, from the Bahamas to the U.S. Virgin Islands, after being purchased by the National Oceanic and Atmospheric Administration in 1977. *Hydrolab* was eventually retired in 1985 and moved to the Smithsonian Institution's National Museum of Natural History in Washington, D.C. During its lifetime, it housed more than 500 researcher aquanauts – one of the most productive habitats ever built.

In February of 1969, as a result of discussions between the National Aeronautics and Space Administration (NASA), the General Electric Corporation and the Office of Naval Research, the *Tektite* I habitat began operations off the island of St. John in the U.S. Virgin Islands. The program was principally the brainchild of Dr. James W. Miller and, because of his persistence and well thought out leverage of the three large bureaucracies, *Tektite* was born. The undersea mission included aquanauts from NASA, the US Navy and the Department of the Interior. The accumulated

THE HYDROLAB HABITAT

interests were crew psychology and interactions; ocean engineering; and diving physiology and techniques. *Tektite* I included a crew component of four aquanauts who remained on the seafloor at a depth of 43 feet for 60 days.

With an unbroken string of success, the United States Navy had made firm commitments to the placement of humans to live and work undersea. To that end, the Navy established what they called an 'Ocean Engineering Test Range' off San Clemente Island, California. Here they planned to test vehicles, equipment and other undersea technology from a manned habitat. They assigned a permanent ship and crew to the range, the USS *Elk River*, and to the newly refurbished SEALAB II that had also been re-designated SEALAB III. SEALAB III had been modified to work at a depth of 600 feet.

With the newly outfitted and improved SEALAB III came a unique program that called for rotating six to eight member aquanaut teams operating for 12 days per mission at 600 feet in depth. The extensively designed program included undersea construction, salvage and all forms of physiological testing. The Navy integrated civilians into the SEALAB III team that were developing a full suite of programs from marine biology to geology and ocean engineering. The prime purpose of the SEALAB program was to boldly extend routine manned operations to nearly all continental shelf depths.

On February 18, 1969, SEALAB III was placed on the seafloor at a depth of 600 feet. Unfortunately, the habitat began to leak and six divers were sent to repair it, but they were unsuccessful. A second team was dispatched and tragically, during the second attempt, aquanaut Barry Cannon died. It was later found that his breathing apparatus was missing baralime, the chemical the team used to remove carbon dioxide. Cannon became the first diver to lose his life in a saturation diving mission. After Cannon's tragic death, the Navy cancelled not only SEALAB III, but closed down the Ocean Engineering Test Range and abruptly discontinued the Navy's aquanaut program. The US Navy, who had been one of the biggest players in the game, abruptly walked off the field and quit altogether.

On April 4, 1970, five aquanauts entered a newly refurbished *Tektite* II habitat at its previous location in the Virgin Islands. In a series of missions ranging from two to four weeks, a total of 53 scientist-aquanauts lived and worked in *Tektite* II, including an all female team. *Tektite* II's final mission was completed on November 6, 1970.

TEKTITE HABITAT

But the United States and France were not the only nations involved in undersea habitation experiments. From 1968 through 1975, other nations around the globe initiated programs including *Chernomor* (USSR: 1968-1974), *Seatopia* (Japan: 1968-1973), *Helgoland* (Germany: 1968-1976) and *La Chalupa* (Puerto Rico: 1972-1975).

Indeed, undersea habitation programs included the big and the small. From the expensive and large to the one man operations, undersea habitation peaked in 1968 when there were fourteen habitat construction and habitation programs underway, as shown in Figure 1 on the next page.

There were approximately as many habitat programs occurring during the single year 1968 than there would be combined over the next 40 years. In fact, from the original habitat in 1962 until the date of this publication, more than 80 percent of all habitat programs occurred in a cluster of activity during the first decade of the aquanaut's history.

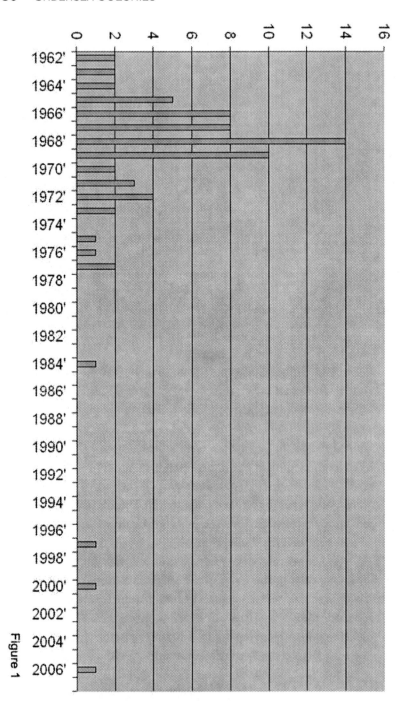

Habitats Constructed Vs. Year

Figure 1

After 1969, the number of undersea habitat programs dropped precipitously until, by the mid-1970's, they had leveled off to less than a handful of programs on any given year. The graph in Figure 1 depicts one of the most astonishing phenomenons in any human exploration initiatives.

Here you can clearly see a wholesale loss of an entire planetary wide exploration activity that stalled in 1968-1969 and then inexplicably collapsed between 1970 and 1973. Please keep this graph from Figure 1 in your mind as we progress though this book. It clearly illustrates a key idea that will be referenced throughout this work.

It is also important to distinguish between single mission habitats and multiple mission habitats. A vast majority of undersea habitats were designed for single missions or a pair of missions, and then were retired.

Considering all the undersea habitats on record, only five were designed for ongoing operations over multiple years: *Hydrolab*, *La Chalupa* (which was later renamed *Jules' Undersea Lodge*), *Medusa* (which was later renamed *MarineLab*), *Aquarius* and *Baylab*. All the other habitats were retired after one or two missions (SEALAB III being cancelled before the mission began).

Further, of all the habitats built, a majority were designed for short mission runs of a fortnight or less (predominantly one week or less), as shown in Figure 2 on the next page. Only nine habitats have hosted missions for more than two consecutive weeks duration, and only three habitats have hosted missions lasting more than 30 consecutive days.

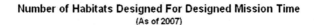

Number of Habitats Designed For Designed Mission Time
(As of 2007)

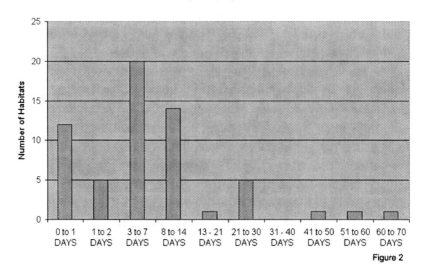

Figure 2

Figure 3[6] below depicts the longevity of undersea habitats. Sixty-five percent of them were only used for a single mission, and less than five percent of all habitats have been used for more than a decade.

Undersea Habitat Use Periods

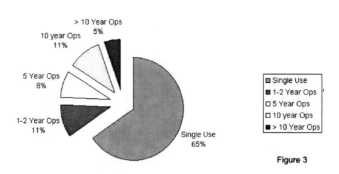

Figure 3

---

[6] The data compiled to construct figures 1-3 was derived from the exhaustive cataloging of Miller and Koblick's essential book, *Living and Working in the Sea* -2nd Edition, plus recent records.

From the early 1970's until 1985, there was only a single door open for aquanauts to access the sea, and it was restricted to marine researchers and scientists. It was NOAA's *Hydrolab* which was retired in 1985.

From 1985 until this writing, undersea habitation has been primarily dominated by three stalwart habitation programs using proven habitat designs. Two of the habitats are located within feet of one another in Key Largo at the Marine Resources Development Foundation (MRDF) lagoon: the *MarineLab* and *Jules' Undersea Lodge.* The third is NOAA's *Aquarius,* located about 15 miles away. Of the three habitats, only the *Jules'* and *MarineLab* are operated continuously year round.   It is interesting that the three most utilized undersea habitats in the world are located in a 15 mile radius of one another. Since the mid 1980's, the history of the aquanaut has been primarily dominated by these three undersea habitation programs.

In 1970, the Marine Resources Development Foundation was established in the U.S. Virgin Islands by the *La Chalupa* designer and mission manager, Ian Koblick.   In 1985, MRDF moved to its permanent headquarters in Key Largo, Florida. In that same year the *MarineLab* habitat was moved to the MRDF lagoon and became available for use to classrooms and researchers. The following year, Koblick and Neil Money purchased the *La Chalupa* habitat and refurbished it into the world's first undersea hotel accessible to anyone with a SCUBA certification – the *Jules'.*

These opportunities provided by Ian Koblick at once signaled the single most important progress in the history of the aquanaut to date. With the genius and persistence of his vision, Koblick literally and single-handedly at last opened the doors to Aquatica to everyone. No longer was the undersea world available only to governments, scientists and researchers with relatively large budgets, but it was now accessible to almost any SCUBA diver.

NASA's initial foray into using the undersea environment as an analog to the space environment was the *Tektite* program in 1969-1970. But the space agency took another look at the undersea world, and the analogous experience of the aquanaut, and in 1992 began a series of observations at MRDF.   This mission was designed and implemented by Mark E. Ward and Richard Presley and supervised by NASA Psychologist, Dr. Al Holland. The purpose of the mission was to simulate the isolation conditions of a long term space mission. It placed aquanauts in an undersea habitat as a space analog for 30 days, and was called the La Chalupa 30 Mission.

Overlapping this cooperative program was a privately funded endeavor, sponsored by MRDF, called Project Atlantis. The program raised public awareness about the possibilities of living and working in the sea and resulted in the establishment of a world record for living in an undersea habitat: 69 continuous days set by aquanaut Richard Presley.

And in the same lagoon, at the time Presley was racking up his 69 continuous days, worked Chris Olstad, the long time manager of the *MarineLab* habitat. While Presley holds the record for continuous unbroken days as an aquanaut, Chris Olstad holds the record for most accumulated days working as an aquanaut undersea. No other aquanaut even comes close to Olstad for logged aquanaut time and someone who works as a professional aquanaut every day on the seafloor, and has continuously since 1984. The only reason Chris is not considered the first Aquatican is that he returns to his land based home on most evenings.

Figure 4 below compares the record holding single undersea habitat mission with the record holding space mission. Rick Presley spent 69 days living and working in the sea compared to Valery Polyakov's 438 day record in space. The record in the supremely difficult to reach environment of space surpasses the lengthiest stay in our oceans by more than six days to one.

Space vs. Undersea Habitation Records

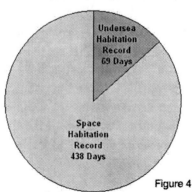

Figure 4

THE AQUARIUS UNDERSEA LABORATORY - NOAA

In 1987, NOAA built the ambient pressure *Aquarius Undersea Laboratory*. Its first operations began in 1988 in the US Virgin Islands. Following damage by hurricane Hugo in 1989, the habitat was refurbished and moved to the Florida Keys off Key Largo. It began its missions there in 1993. *Aquarius* was retrieved for refurbishing in 1996 and operations resumed in 1998. Its base-plate rests in approximately 62 feet of water, with the habitat mounted off the bottom at a depth of approximately 47 feet (tidal range at the site is between two and three feet). At the end of a mission, aquanauts undergo a 17-hour decompression that is conducted within *Aquarius* itself, while on the bottom. After decompression, the aquanauts exit *Aquarius* and SCUBA dive back to the surface.[i]

In 1994, NASA utilized the *MarineLab* habitat, employing the undersea environment to simulate the space environment, in what was called the OCEAN Project, an acronym for 'Ocean CELSS Experimental Analog – NASA'. For the first time, an agriculture crop (lettuce) was planted and harvested in a habitat on the seafloor using a hydroponics system in a simulation of a bioregenerative space system employing equipment and techniques developed for space agriculture.

In 1997 and 1998, NASA returned to MRDF with its own habitat, the *Scott Carpenter Space Analog Station* (SCSAS). The SCSAS was designed to be launched from a trailer like a boat, towed to the MRDF site and lowered to the seafloor between the *Jules'* and *MarineLab* habitats. In 1997, the habitat's 11 day mission was designed to run concurrently with Space Shuttle Mission STS-86 while on its mission to the Mir Space Station. In 1998, the purpose was to integrate with John Glenn's Space Shuttle Mission STS-95 and act as an education and outreach mission. During both years, the station connected with classrooms, museums and aquariums all over the world by telephone and Internet links, acting as a space analog model.

AQUANAUT
MARK E. WARD

In 2000 Mark E. Ward was contracted by The University of North Carolina, Wilmington (UNCW) and the National Undersea Research Center (NURC) to develop and coordinate media and public outreach activities and strategic partnership opportunities for *Aquarius*. In the spring of that year, Ward in turn contacted colleague Bill Todd, a Simulator Supervisor with United Space Alliance in Houston, and invited him to visit the habitat near Key Largo, Florida, with a view toward exploring possible collaborations between NOAA and NASA. Soon thereafter plans were developed to conduct an initial test mission using the *Aquarius* habitat.

The mission took place in October 2000 and was dubbed SEA TEST. As part of this mission, Bill Todd saturated for several days onboard *Aquarius*. Following this test mission, and due to the persistent efforts of Mark Ward, Astronaut-Aquanaut Mike Gernhardt and Bill Todd, NASA management was convinced to fund a follow-up mission, and NEEMO (NASA Extreme Environment Mission Operation) was born. NEEMO has played a key role in meeting the requirements for testing NASA astronauts' crew psychology, training, and equipment trials in a remote and extreme environment.   Ward continued to support NURC's interests and Todd's efforts through 2004. As of May 2007, 12 NEEMO missions have been undertaken.

The *Aquarius* program was also introduced to key personnel with the US Navy's Experimental Dive Unit (NEDU) in Panama City, Florida, by Ward. His efforts led to increased cooperation between the Navy and NOAA, sharing of resources and personnel, and several collaborative undersea missions in *Aquarius*.

In addition to helping establish these two strategic partnerships for NURC, Ward also introduced *Aquarius* to dozens of science centers, aquaria and museums, delivering presentations at public gatherings and coordinating live interactive broadcasts that significantly heightened public awareness of undersea habitation. He also worked with scores of media outlets and other PR interests throughout the course of his association with *Aquarius*.

In 2000, SEALAB, *Tektite* and *Hydrolab* veteran Morgan Wells deployed the ingeniously designed habitat, *Baylab*, which began operations in Chesapeake Bay. Wells' genius was in his use of a unique, low energy life support system and the integration of a one-of-a-kind end hatch, rather than the traditional moonpool approach with an opening on the bottom of the habitat. In *Baylab*, aquanauts simply enter through a hatch positioned relatively low on the end of the cylinder, rather than through the bottom, stand up and step over a sill into the wet room! This allows *Baylab* to be deployed nearly on the bottom.

Baylab also features the capability of operations as the only completely self contained habitat ever built with no connection to the surface, if desired. It also features a unique ultrasonic surface communications capability that enables the habitat to operate completely independent from any surface connectivity at all during its mission. *Baylab* is still operational.

In 2007, Lloyd Godson, the winner of *Australian Geographic's* 'Live Your Dream' competition, created his *BioSUB* from a simple steel box to study the use of linking an active algae system into the life support system of an underwater habitat. That

12 day mission began on April 5th, 2007 and ended successfully on April 17th. It was the first time that a bioregenerative component – an algae system – was used as a significant part of an undersea habitat's life support system.

The mission was so popular that Lloyd Godson is reprising his fantastically successful and interesting experiment in Australia and is

AQUANAUT LLOYD GODSON

building a follow-up model to his original *BioSUB*; a *BioSUB* 2. Current plans are to place the new habitat on the Great Barrier Reef. Lloyd is building the new habitat as the centerpiece of a television series to be produced and broadcast by a Canadian television network.

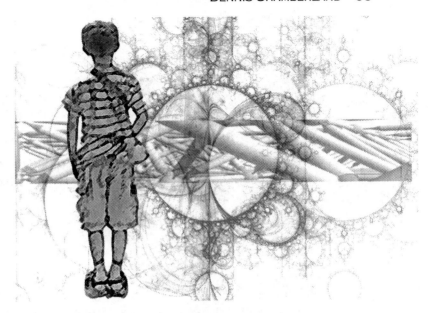

THE GREAT ABANDONMENT

"If we are content to live in the past, then we have no future,
and today is the future."
Louis L'Amour

While it is true that exploration of the undersea regions has been continued by a dedicated group of aquanauts from 1970 until the present, it is equally true that those efforts can be characterized as sparse, at best. Today it is almost wholly driven by individuals, rather poorly funded government groups and well known project organizations like the Cousteau ventures. And were it not for the single-handed, nearly heroic, individual effort of Ian Koblick, the door to the undersea world would have been slammed shut nearly altogether.

As you look at the data, it is obvious that near 1970 the enthusiasm in the world for undersea habitation did not just wane, it, in fact, altogether crashed in what can best be characterized as a great abandonment.

There were reasons aplenty – perhaps better stated as excuses – for this abandonment and each set of unique motives can be traced to the individual endeavors. From 1969 to 1971 there was a total abandonment by two of the biggest players in the field who abruptly threw in the towel and walked away: the US Navy and the Cousteau team. Yet, as statistics clearly demonstrate, the small, single ventures also ceased at nearly the same moment in time. It is more than just an historic coincidence that cancellation occurred nearly everywhere simultaneously, as will be discussed in the next chapter.

In 1968, the US Navy had established a large undersea test range off San Clemente Island specifically for the purpose of establishing undersea laboratories for the suite of equipment that would be managed from a manned undersea base. But, after the tragic death of diver Berry Cannon, the Navy quickly yanked the Sealab III habitat, shut down the test range, sent the support ship home and walked away, washing its hands of any mission involving men living and working from fixed bases in the sea.

Three years before that, Conshelf III had ended and Cousteau was now off sailing the world's seas on his *Calypso* and making stunning documentaries and books, but with absolute silence in terms of any hope for a Conshelf IV mission. It was as though, with Conshelf III, he had answered his own question about man living and working in a habitat on the seafloor. But whatever that answer may have been, Cousteau was not sharing it widely.

By the early 1970's, the whole world of exploration had abruptly and radically changed. One morning, the planet awoke to an entirely new language of exploration, both in the oceans and in space. The alteration of expectations seemed sweeping and took place almost overnight, with the profound effects reverberating throughout the world of exploration much like a Tower of Babel experience. The great races were over and the crowds had gone home. The United States and the Soviet Union had ceased reaching for the moon and beyond, settling quite contentedly into low earth orbit. And in the seas, the two key players had abandoned the idea of permanence and had gone quietly on to other things. It was as though they had discovered that man was not meant to live and work in the sea.

There are many speculations on each of these instances and they will most surely remain speculations. In the case of the Navy, the ultimate reasons were probably classified and now lost. And in the case of Cousteau, his decisions may have well been personal. There is no way to really know for sure.

However, there are broad hints. The Navy was embroiled in an unpopular and expensive war in Southeast Asia that was straining its every asset. The SEALAB III habitat was plagued with technical problems and there are well placed rumors that there was even a saboteur at work in the group. It seems likely that at some hierarchy in the Navy's command structure, the death of Barry Cannon was the final straw that made the effort appear at least on the fringe of, if not altogether outside, the Navy's primary mission. The plug was unceremoniously pulled. The questions being asked had become larger than the resources the Navy was willing to provide to answer them during a time of serious national conflict.

At the same time, the statistics aptly demonstrate that just as the big players were backing out, so were the small habitat builders and small team aquanauts. It was as though there were some wall that had been encountered and that, almost like a switch, the rush into the oceans had suddenly ceased by all players, small and large.

In the case of Cousteau, there are sources that seem to provide a stunning and direct answer to the issue of why there was never a Conshelf IV. Here are two quotes that may offer a surprising hint as to why Jacques Cousteau pulled Conshelf III off the bottom and never returned:

"Conshelf III ended the underwater living experience for Cousteau and company, now heavily in debt from the project. By 1980, Cousteau would rescind his long held belief that man would one day live on the ocean floor by declaring colonization of the sea to be unrealistic."[a]

And in another reference:

"In the late 1960's, Cousteau found new challenges and opportunities that made him change his course. He turned from doing scientific research to popularizing it. Fifteen years later he would utterly condemn the idea of colonizing the sea."[b]

There are several conjectures that can be drawn from these comments, although they are admittedly sheer speculation. It is possible that a Conshelf IV would have cost more than Cousteau had to spend. It is possible that Cousteau felt he was at a fork in his career path and that in his mid-50's he could have selected either a Conshelf IV or a sweeping set of well funded, world ranging televised expeditions, and he chose the latter to leave as his greatest legacy – but he could not manage nor afford both.

We will never know for certain, of course, and any further speculation would not only be unethical, but tarnish the memory of a single man who created and accomplished more to educate everyone and spur their interest in the undersea world than any other single individual.

Yet standing out like a flashing neon sign in these comments is Cousteau's alleged, blunt repudiation in 1980 of all that he had accomplished to open the sea for colonization mentioned above.

If true, it is a stunning admission from the mouth of the man who declared Aquatica to be the next great frontier, and even christened his missions after the continental shelf areas that he claimed man would one day permanently settle. It is something akin to Werner Von Braun declaring the colonization of space or exploration of the moon to be 'unrealistic'. It embeds, at once, the disbelief encompassed by the loss of a dream and the simultaneous turning of a hero's leadership to the hero's abandonment of the dream he inspired in so many others. To say such an admission is shocking is a gross understatement.

And yet perhaps to us, decades removed, it should not be so much a disappointment as it is a signal. Does Cousteau call to us from his grave and beg yet another look at his decision? All excuses and useless speculation aside, it is very possible that

Jacques-Yves Cousteau was right on target and correct, as was the United States Navy. In the end, it is not only possible, but highly probable, that all parties made the right call at that time after all, when we examine exactly what they, in fact, discovered on the bottom of the sea and why they chose not to return.

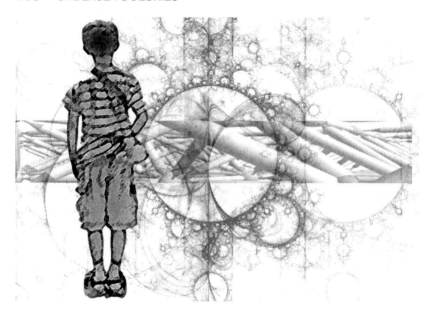

THE MISDIRECTED START IN UNDERSEA SETTLEMENT

"Success is not measured by what a man accomplishes,
but by the opposition he has overcome,
and the courage with which he has maintained
the struggle against overwhelming odds."
Orison Swett Marden

Jacques Cousteau has always been, and will always be, one of my most admired heroes. I thoroughly relate to one story of him as a boy. Cousteau loved motion picture cameras so his parents purchased one for him, and young Jacques went after his hobby with a single-minded and voracious passion. Unfortunately, it also resulted in his ignoring his studies and his beloved camera being impounded as punishment until his grades improved. Jacques was so upset by the decision that he broke out 17 windows in his parents' Paris apartment. This action resulted in him being sent to a boarding school to learn discipline and academics, while giving his parents a much needed rest.

I can completely understand this story. Here was an individual who responded to life's challenges with a single-minded determination to see his goals tough. It is a window into his heart that I recognize and share in my own. And so it is that I can also understand Cousteau's statement about undersea settlement being 'unrealistic'. But that reference, when taken out of context, can, and has, led to a gross misunderstanding of what I believe he was truly trying to say.

Did Cousteau actually 'repudiate' the whole idea of colonizing the sea? I can only find a single source for that comment in an unauthorized biography. Unfortunately, from that single, admittedly unauthorized source, the comment has been repeated innumerable times. But if I may take an author's license and paraphrase his comment, please allow me this liberty. I believe that Jacques Cousteau was trying to say this:

"Undersea settlement is unrealistic, *given the approach we took.*"

Understanding a misdirected start in exploration requires a dual understanding of the merged human disciplines of psychology and sociology, for both of them work together in initiating a misdirected start, and then in perpetuating it.

Every misdirected start in exploration leads to a dead end. Once the dead end is reached, the explorer has to back out and start over again given the lessons learned. This process requires two disciplines: flexible wisdom and humility. Without both of them working together, it is impossible to back out and start again. Otherwise, the stubborn explorer stays and bangs his head against the obstacle until he dies, or he is forced to back out, regroup and make a decision. That decision is: do I go back and try this again or do I stop and figure out another way?

Determining that a misdirected path in exploration exists at all is manifestly difficult and generally has historically required that all the old explorers die off before the error is remedied. This is because the average explorer's mind is not normally equipped for much humility, and his wisdom can be, on occasion, fixed and difficult to reprogram. Explorers require bravado and an overly inflated sense of self importance, such that flexibility is rarely a part of the package (I, unfortunately, speak with much experience). These traits, although they may sound negative, are the precise characteristics that make explorers what they are, and enable them to do what they do so well.

Admitting a misdirected path in exploration directly implies that false assumptions were made and the path to discovery is therefore going to be severely impeded to the point, as Cousteau properly stated, that it becomes 'unrealistic' to actually reach the objective. I believe the Navy and Cousteau discovered this truth simultaneously. Instead of sticking around to back up and solve the misdirection, Cousteau went on to fulfill his life and leave a different legacy and the Navy went on to focus on their prime mission. In retrospect, both seem like sound decisions when the misdirected start had been discovered.

We are all masters of misdirected starts. From the first day of our existence, we are all creatures who move about, make decisions and formulate our plans based on, of all things, misdirected starts we have made in life's journey. From the first clumsy attempts to walk that culminate in eating Dad's shoes, we learn that some paths are profitable and others are most certainly not.

For instance, when I was a Minorities Student Advisor in the Engineering College at Oklahoma State University, I learned that the average undergraduate student changes their major three times before graduating. Two of the alterations resulted from misdirected starts in the student's educational path. But they were an accepted part of the educational experience. Certainly, no fool would have condemned the students as abject failures just for changing their majors to a field they discovered they preferred!

In personal relationships, misdirected starts are far more common than forging lifelong bonds at the first meeting.

Statistics in the United States demonstrate that the average worker changes career pathways from nine to 13 times in a working lifetime.[a]   These changes are for various reasons, but many of them are misdirected starts: in career choices, company decisions and other misdirected career pathways.

In science, the concept of misdirected starts is actually engineered into the scientific method!   Misdirected starts are considered a legitimate part of the process itself. Indeed, were it not for misdirected starts, no progress could ever be made!

The number of misdirected starts in exploration is legendary. Just to name a few: the misdirected starts involved in the great Northwest Passage enterprise; both polar explorations; and most great mountaineering exploits.

Space Exploration is even now involved in reconciling a great misdirected start of its own, which relates directly to the misdirected start in undersea exploration, for both were born out of the same culture and both inherited the identical and altogether false expectation. It all began with the culture – the nameless culture that both inspired and motivated all the men who were responsible for the world's greatest period of human exploration in the 20th century. The culture had just defeated the enemies of freedom and unrestricted capitalism. Literally, the message was unambiguously communicated, by the strength of our free culture, "If we can dream it, we can do it."

Hinged to that false expectation was another just like it: "Mankind has a nearly unlimited adaptability. And not only is he adaptable, he is strong and he can adjust." On these two expectations, the road to all the misdirected starts was paved.

As we have discovered both in space and in the oceans, man is certainly not as adaptable as we once thought he was. To the contrary, we have learned that man has a sharply limited capacity to adjust to every environmental condition. Unfortunately, the expectation was freely communicated that man could adapt to various high pressure environments that would allow divers to interface with the oceans directly through an always open moonpool no matter what the depth. In space, we expected that microgravity would be a never ending lark and that we could fly all over our spaceships whenever we wanted.

Both expectations proved to be misdirected starts. Yet today many people steadfastly cling to those old expectations. Missions are still designed around them, and we still plan with the idea in mind that someday, somehow, humans will adapt if we can just figure out that ideal countermeasure.

The mental process at play is to demand that the plan work, come hell or high water, to preserve the initial expectation that man can endlessly adapt. The solution must therefore only be one countermeasure away: the right exotic gas mix or the electronic speech un-garbler or the pill that fixes aseptic bone necrosis in the oceans or chronic cardio atrophy and permanent bone loss in space.

But these examples are simply inconveniences compared to the more fundamental health effects and issues of living and functioning in hyperbaric environments. Relatively recent studies have demonstrated serious health problems associated with hyperbaric environments. In one study it was determined that male fertility was 'seriously compromised' in a hyperbaric environment at a relatively shallow depth of 150 feet during a 33 day mission. The study reported that following the dive, infertility persisted for months – up to day 82.[b]

Other effects of living in hyperbaric atmospheres include, "… airways dysfunction and changes in lymphocyte population numbers leading to a relative immunosuppression. Changes in neuropsychological function have also been detected, including altered memory and cognitive ability with corresponding brain stem reflex abnormalities and EEG changes."[c]

Prolonged exposures to hyperbaric environments have profound, negative effects on human bodies. The effects are so intense that if humans wanted to live in an ambient pressure colony exposed to the outside pressure at a depth as shallow as 150 feet, they would not even make it beyond the first generation, because, even if they managed to fight off all the other physiological challenges, the men would become sterile and never be able to reproduce!

It is endlessly fascinating to me how most humans cannot step back, humbly admit they have made an error in estimating the exploration process and then flexibly retool the next generation of systems to meet what they have discovered. After all, is not 'discovery' the ultimate purpose of exploration? Instead, they cling to old notions that the first generation of explorers found to be roadblocks, to continue with the dream.

For example, which is the more difficult process: inventing 5000+ individual engineering and medical countermeasures to microgravity   or   engineering   the   single   artificial   gravity countermeasure by simply spinning the spacecraft? And in the oceans, what prevents us from simply placing a pressure hatch between the open moonpool and the living quarters so the aquanauts can live and work at near one atmosphere?

The human body is designed to work at one-normal earth conditions, defined as one atmosphere with the component consistency of air and one gravity at room temperature. We have discovered that the body will rally and make the attempt to adapt to many kinds of environments, but that adaptation process is neither easy nor is it healthy. In nearly every case, the human body suffers, and often permanently.

The misdirected start in undersea habitation all began, of course, with the expectations of George Bond. Yet, obviously, Dr. Bond commenced his work before any of this was known. He was asking the question – not making the claim! He is not to be blamed for the misdirected start; but instead it is important to see from where it originated.

George Bond's hypothesis was paramount in his mind: that man could adapt and work effectively in pressure. He was correct. Man could adapt and work effectively in pressure. But in the process he discovered that he could do so for only relatively brief periods of time; and all the while, the body was changing, and not necessarily for the better. But from his writing it seems obvious that Bond was also capable of separating the goal from the mechanisms that enabled the goal. Hence I am convinced that Dr. Bond would himself utilize the lessons learned – back up and start again with the new information his work, in fact, enabled – and still reach his original goal!

In the oceans, Jacques Cousteau and the Navy both gave broad hints to intractable problems long before they shut their programs down. The details that emerged pointed to issues with hyperbaric gasses such as: chronic leakage of helium into and out of every human built device; cold and thermal issues; unintelligible speech; and long, post-mission decompression times. And yet neither of them admitted to a misdirected start or recommended one atmospheric seafloor systems with diver lockouts. Cousteau seemed to be on the right track with the dive saucer accessibility solution, but it never came together as a coherent undersea engineering plan.

One thing is certain; Cousteau voiced clearly and accurately the truth of his unambiguous discovery, with my added caveat:

"Undersea settlement is unrealistic, *given the approach we took.*"

Jacques Cousteau and the US Navy both arrived at the dead end of the first tunnel at the same time and called the shots as they were – *and they were both entirely correct.*

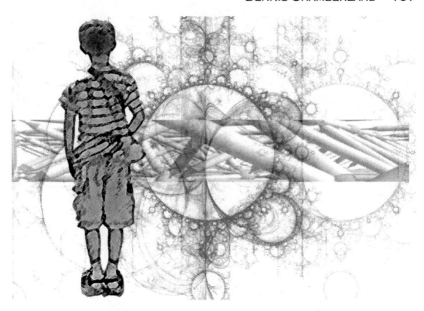

THE NEW PATHWAY TO PERMANENT UNDERSEA SETTLEMENT

"Failure is success if we learn from it."
Malcolm S. Forbes

T he first pathway into Aquatica led to a dead end. Venturing further, deeper or even investing more time and resources in this approach was far too risky, both technologically and physiologically. The previous experiments that led up to a dead end represented a misdirected start in undersea exploration.

But this does not, by any means whatsoever, mean that mankind will never permanently occupy the undersea world! In fact, it only means that the pathway to that end has simply been altered, and not even significantly so. Now there is a new path and 21st century aquanauts will utilize many of the older technologies as they follow a fresh track into the sea.

It also does not mean that there will never be a requirement for saturation diving in the future. Indeed, occupational and military divers will utilize this technology indefinitely thanks entirely to the pioneering work of Bond and Cousteau. But it will be no less plagued with problems and risk than in the past. And it is definitely, as Cousteau rightly expressed, an 'impractical' approach to long term undersea settlement for entire colonies of humans.

I am by no means a prophet, but the following few pages will present what may be the new pathway to the original goal of George Bond and Jacques Cousteau.

Terms immediately crop up as the first impediment to understanding. Dr. Bond described his program as 'saturation diving'. Implicit in this phrase was the perceived requirement for decompression before the diver or aquanaut could return to the surface. However, an aquanaut may be completely 'saturated' with various diving gasses (including nitrogen – a prime gas of interest) under 21 feet of salt water (FSW) or less and still not have to decompress upon returning to the surface. Hence, in any new approach, it is essential that the student of undersea habitation understand the terms and how they are used. If an aquanaut enters into a habitat or a habitat compartment pressurized a single pound per square inch over sea level, a process of 'saturation' begins.

In the new approach, instead of requiring physiological adaptation, the aquanaut will be protected against the need for decompression or any decompression events. Instead of demanding the human adapt to the system limitations, the system will be required to adapt to the human limitations. This foundational mindset is so critical that I have stated it as my First Law of Human Exploration and repeated it again as the First Principle of Undersea Colonization. It is a philosophy not only of system design but of exploration design as well. It satisfies the argument that the human will be protected as it refutes the argument that the human requires no protection and can adapt.

In the new approach, access to the undersea environment will no longer be direct. If the aquanaut wants to enter Aquatica, he will have to pass though a pressure lock-out or an undersea airlock dive chamber before he can drop into the open sea. Likewise, if the aquanaut enters the open ocean environment, he will be required to monitor his decompression limits before reentering the depressurized portion of the habitat.

The new approach does not necessarily do away with what we have learned about mixed gas diving. Certainly a mixed gas diving capacity will always be present onboard the habitat to allow for more extended work outside the habitat structure. But the new approach would require the aquanaut to use a hard suit or a mini-submarine to perform extended work in deep waters outside the habitat.

Each of these limits may seem to reduce the capacity of the aquanaut when compared to the old system called 'saturation diving'. Yet, when you compare the total system requirements, system cost, hazards, personnel demands and time for end-of-mission decompression, the new pathway is actually less costly.

Under the old system of exploration, the demand and overt expectation was that the human be directly exposed to the environment with the assumption of adaptability. This philosophy is so deeply engrained in many people that even discussing it can easily lead to violent disagreements. Yet, if the exact same expectation of direct exposure were to be levied on a space system, the proponent would be considered insane.

For instance, how logical is it to expose an astronaut to the vacuum of space and then demand that the astronaut 'adapt'? Or how realistic is it to expect that the hatch to a spacecraft be left open to the ambient pressure of space just to make it easier to access the environment? If it makes perfect sense to give the astronaut protection against the space environment – in terms of both temperature and pressure – then why does this expectation seem so alien to the undersea explorer?

The answer is, of course, the original expectation has not been lost to the previous generation of explorers. Old ideas and old notions, even when conclusively proven not to work, die very hard. As was stated before, they typically require the older generation to die off before the new, more realistic and effective ideas born of the lessons from the past take root.

It is very likely that many of the new generation of undersea explorers will not even 'dive' to and from their habitats. They will almost certainly ride down in diving bell type 'elevator cars' or in submarines of various sizes. The new aquanaut will probably rarely even get wet on his way to and from the surface. They will enter into habitat compartments that are pressurized to 1.6 atmospheres (24 psi) or less. When working outside for periods approaching decompression limits, they will either wear hardsuits or be in mini-subs using robotic hands.

The best example of this approach is found in submarine technology. As was clearly discussed previously, a submariner is not an aquanaut nor is a submarine an undersea habitat. I only invoke this discussion to make a point.

The submariner enters into his submarine at surface pressure. He closes the hatch and rides undersea at various depths, returns to the surface, pops the hatch and goes about the surface wherever and whenever he wishes without restriction. This plan has worked flawlessly for more than 200 years. Why? *Because the human limit was never challenged.* The technological expectation was always one atmosphere at comfortable conditions. Hence, there were never any challenges to the technology expanding and moving forward to the point where we are today, with crews of thousands of men in undersea missions all over the world on any given day. It works for one simple but very powerful reason: the human is protected.

These submariners are by no means permanent settlers in permanent undersea communities, but they are there living and working safely and very successfully. The key to their success is that they do not challenge the human capacity to adapt to dangerous and usually miserable levels. The expectation has always been that the submarine technology adapt to the human limits and not the other way around.

I recently had this very discussion with an aquanaut friend. He responded that if things went this way, he would miss 'instant and unimpeded access' to the sea through his open moonpool and that he would also miss the swim from and to the surface. It occurred to me that our discussion was like trying to discuss the concept of an elevator and an automobile to a farmer living in an adobe cabin on the prairie. The farmer would undoubtedly miss his uncovered horse, even in the rain, and would prefer it over an automobile.   And he had certainly never used an elevator, nor would he have any use for one if he had it.

The aquanaut was a diver and loved to dive. Likewise, the farmer was a horseman and loved to ride. But the key idea was that the farmer and the aquanaut's sentiments were not based on anything more than their feelings about the life they had gotten used to.

The deep system aquanaut would have a different opinion, of course. While he may not miss the constant cold, pesky helium and the squeaky voice, he may balk at donning a hardsuit or the hassle of using robot arms on the end of his mini-submarine. But in the end, even his arguments are based on what he has been used to, not practical limitations of the new systems over the exposed and vulnerable human being.

In the new pathway that will lead to permanent undersea settlement, the human package will be protected as a matter of design priority. All other systems will follow suit as the new age of underwater exploration and settlement begins.

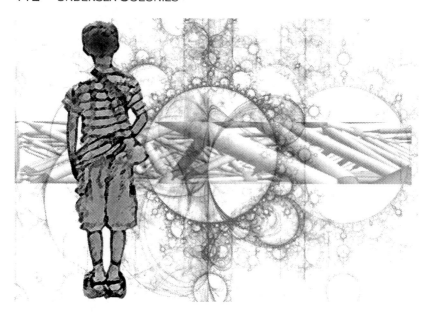

WHO REALLY WANTS TO GO THERE TO LIVE AND WHY?

"Every noble work is at first thought impossible."
Thomas Carlyle

I have personally met many sane, reasonably intelligent and very motivated people that, if there was a community to go to, would live out their lives as permanent residents of the undersea dominion of Aquatica. And they would do it today, with great enthusiasm. None of them are kooks; they are not weird nor are they loners. They are the ones who are even today being called as the first Aquaticans – as permanent human citizens of the undersea kingdom that will be.

The question 'why' inevitably pops up first. In our past history, humans have gone down to the sea and have drowned. Or humans have gone to the sea and disappeared over the horizon and they never came back. The idea that a sane human would go down to the sea and purposefully live under its surface is invariably connected with unreasonable danger and imminent death – therefore it is also connected with peculiar people. But that is a myth. That image is linked to watching way too many movies and an overactive Blockbuster video card.

While it is true that some people go to the sea and die or disappear, the numbers are, in fact, far lower than of those humans who go out to the local highway and die or disappear. No one considers anyone else a kook or strange who gets into their car and drives down the dangerous highways on land. And yet, on this very day, there are those who will die there; and some will vanish, never to be seen again. But we do not consider that strange. It is but a sad fact of everyday life; a probability; a toss of the dice.

Yet, that same highway, while offering up tragedy, mostly does not. Through careful engineering and laws built on past tragedy, we have learned how to make highways the very arteries of life for our culture. While there is some danger there, we are willing to take a chance every day in order to enjoy the highway's vast benefits.

I have sat down with many people and purposefully had the discussion of living permanently undersea. Many of these discussions were held in undersea habitats while we sat in warm, dry, safety, living and working in the very place where we wanted to settle permanently. We knew as we sat there that our current mission would soon end, but none of us really wanted to leave – and that engendered the discussion. The reasons expressed by these aquanauts for living permanently in Aquatica are many, so it is impossible to say that everyone who shares this dream is alike.

Some aquanauts wish to live in the warm waters of the tropical sea. Others wish to live suspended in the void in the middle of the ocean where both the bottom and the surface are unseen. A few wish to live in the total darkness of the deep ocean where the only light is from passing bioluminescence. But they all share the same dream – to live permanently under the waters of the world's seas, in the region of the earth called Aquatica.

If he is very honest, the future Aquatican would admit that he carries two lists in his mind that justify why he wants to live undersea. The first list is very personal. It represents his core, deepest reasons. They are the profound, personal, emotionally driven reasons that actually provide all the true personal energy to want to do this extraordinary thing.

The second list is what I call the 'Sir Edmund Hillary list'. It is the list that is used to prove to everyone else that your dream is not really as bizarre as it first sounds to the novice who has no idea what an aquanaut even is, much less an Aquatican. I call it the Sir Edmund Hillary list because he taught me a very important lesson about exploring and how to link the powerful emotions that drive an explorer to the rest of the world – which tends to be absolutely clueless about these things.

SIR EDMUND HILLARY

Hillary said, "Nobody climbs mountains for scientific reasons. Science is used to raise money for the expeditions, but you really climb for the hell of it."

So allow me be candid and open up both my lists so that you can see clearly what is inside my mind and heart. I take this emotional and professional risk with you, dear reader, knowing that out of every hundred or so pair of eyes that see these words, there is actually a future citizen of Aquatica reading. You may be a pre-teen boy or girl; you may live in the open plains of America or in Paris or in the outback of Australia; you may be of retirement age or in the middle of a successful career. But as you read, you know who you are. And as you read this, you feel the energy welling up from inside; from that unknown force given by the Creator that drives you forward and onward. It has nothing to do with the Sir Edmund Hillary list, but everything to do with the first list and how you are wired.

This chapter is truly a risk for me, but I take it for the sake of those who are wired to understand. It is a risk because some may say it implies insincerity or a hidden agenda. And yet, I hope that it has the opposite effect; for, by opening up all that is in my heart, I cannot be accused of hiding anything. It is one of my purposes for this chapter to show every card in my hand.

You see, if I only had my Sir Edmund Hillary list, I would never be able to pursue the dream. I would not have enough emotional energy or wherewithal to write this book or strive beyond these pages to work for permanent human undersea colonization. My Hillary list is not strong enough to compel this profoundly difficult process. The Hillary list is not, and can never be, wired deeply enough into my inner being to drive the incredible amount of energy and thought required to make any of this happen – because it is that core energy that makes all this work.

I have described before my major driving force to do this as the splinter in my mind. It is that nameless compulsion that relentlessly pushes, nags and drives me on. I cannot describe it any better because it has no name. It is that energy that seeks expression, as the young Cousteau found when it led him to break 17 windows in his parents' apartment. He could not describe it either and, as a lad, it led him to convey his pent-up fire with an unfortunate demonstration. But as I read his story, I immediately recognized Cousteau's energy. It is common in those who have been so blessed with the inner fire.

This inner fire invariably leads to expression. But like young Cousteau and the hapless character, Roy Neary, in the movie Close Encounters of the Third Kind, sometimes it drives one to articulate the energy in what appears to others to be strange actions. Yet to the relentlessly driven, it is not strange at all! Sir Edmund Hillary knew well that he was driven by some nameless force inside that he would never be able to rationally describe. And so he called it 'the force that drove him to climb mountains for the hell of it.' But to another explorer, there is instant understanding of his intent. He climbed mountains because he had no other choice – he had to climb. He would have gone patently insane unless he climbed.

That nameless energy is the prime force that drives me to take part in this activity. But there are other items on my personal list as well. While it is true that I wanted to live in Aquatica long before my first day as an aquanaut in October 1993, that experience only confirmed that the previous drive was real.

That October day was cloudy above the water, so that the seafloor was somewhat dim. I sat inside a large, clear acrylic sphere on the seafloor in blue shorts, white t-shirt and white socks, with a pillow behind my back. I heard the soft hiss of the air filling the sphere and felt at perfect peace. I watched the fish slowly circle around the sphere and, with hardly any imagination at all, I canceled the transparent walls out with my mind and sat suspended in the water with them. The thought struck me immediately: I was not a stranger in a strange land at all. I belonged here. I was a citizen here and I had always been!

How do you ever explain that form of perfect confirmation to someone else? It is an emotional energy that somehow morphs into an invisible connection with a reality inside the closed confines of mind and heart that most people living on the lands will never come to understand. Sir Edmund Hillary gave up and made two lists: one for himself and another, more official list for the non-explorer. To Hillary, he knew exactly what he was doing and why. There were just no human words to explain it to someone else who was not equipped to understand.

As I have spent more than a month living and working undersea, I have come to love it for many reasons. I am attracted to it for a whole list of reasons that I will attempt to categorize for you here.

Some people love living in the mountains. Claudia and I have a beautiful place in the mountains of Tennessee, on the southern tip of the Appalachians, that we call Stonebrooke. It is stunningly beautiful to me. Some people love living on the beach, as do I. Many have dreamed for a place on the shore in the Florida Keys or their own private island in the Bahamas. Some people love the deserts and many have fantasized about a home in the middle of a desert. Other people love living on boats and cruising from one point to another. Many friends of mine have expressed a dream of owning a large cruising sailboat. Some love life on a farm. Others feel most at home on a sweeping, vast prairie.

The point is: people have dreams of where they want to live, and their reasons are complex and tied to deep emotional motivations even they may never understand fully. And yet, for the most part, each human understands this and fully respects the other's personal choices. The same process is true for Aquaticans.

I adore the pace of undersea living. It is generally much slower than on land. I love the idea that my swimming pool is the entire ocean and that I never have to surface to make use of it. I very much enjoy weightless flight from one point to another undersea. I appreciate the wildlife and watching them endlessly. I treasure the colors of the day and the onset of the morning and evening. I take great pleasure in complex habitat engineering arrangements and life support systems. I relish the night sounds of my mechanical cocoon that inevitably lead me to the deepest and most profound sleep I have ever experienced. I am intrigued by the transport systems, the logistics, the communications and the potential for so much more. There is hardly anything I do not love about undersea living. But you can see from this list that these are principally emotional drivers. They are all difficult to put your finger on and to describe. And yet they all combine together to compromise the real energy behind my drive.

In the end, there is a single reason that trumps all others ashore. Land dwellers use it every day and it is universally accepted as 'good enough'. In the free nations of the earth, it is acceptable to say that we go and live anywhere we please for this simple reason: it is what we want to do. Just wanting to do what drives us is understandable and respected as an individual choice in a free society. Indeed, that is what personal freedom is all about – we are not compelled to explain our choices to anyone. We do things simply because we want to, and that is sufficient!

Some people are hunters, some are golfers, some like boats and fishing, some are SCUBA enthusiasts, some like baseball, others like all sports, but some just like professional football. Some people climb mountains and others are weekend campers while still others are dedicated bird watchers. But every person makes their choices for reasons they may not be able to explain perfectly to someone else. The best thing about a free society is that you do not always have to explain yourself, and everyone else respects your freedom to choose. It is just understood that people are different and enjoy different things without having to justify your decisions every step of the way.

Yet, having defended the core energy that drives the explorer to explore, it is important to recognize that while Sir Edmund Hillary had two lists, he was intelligent enough to understand that he actually *needed* two lists! It was just not good enough to defend his choice to climb mountains by stating that he was doing it for the hell of it if he needed someone else's help to do it. And that is exactly why he had two lists.

The Sir Edmund Hillary list is most certainly not a disingenuous ruse. It is the intellectual equivalent of an explorer's Rosetta Stone. It seeks to tie together the energy that drives the task with a second list of valid reasons that everyone else can understand, agree are important, get behind and support together. From the supporter's viewpoint, it is essential to have the program led by someone who is internally driven for sound, rational reasons. The explorer and the supporter strike out together with a common set of agreed upon objectives. Two lists serve two essential proposes: one drives the explorer and the second drives both the explorer and his supporters.

My Sir Edmund Hillary list is found here: eight reasons why it is essential that humans permanently occupy the undersea regions of the earth. In this list, we can all agree that there are causes for going undersea and living there that far outweigh any personal lists or motives. These are purposes and objectives that affect mankind and transcend any one individual or their personal motivations:

Reason Number 1

We are the stewards of our planet. We are morally obligated to know what is happening to the largest region of earth. Pollution and mankind's effect on the environment demand nothing less than that we place our scientists in the ocean full time and discover what is going on there by first hand observation. No machine, no robot and no remote camera have the intellect and capacity to judge as one single, well trained human mind watching *in situ*. We know more about the surface of Venus and Mars than we know about the details of our own oceans. Not having humans in the oceans when we have the technology to keep them there is grossly irresponsible on the part of humanity. It demonstrates our lack of commitment to the responsibilities we have to monitor, safeguard and protect the largest region of our planet. It is also crucial to recognize that our liquid mantle has great impact on the health of the earth and is the area with our greatest biological diversity.

Reason Number 2

Mankind's manifest genetic destiny is, has always been, and will always be, dominion. He has occupied the land. He is preparing to occupy space. And he will soon occupy the oceans. Whether it is accomplished by American explorers or explorers from another nation, the oceans' permanent human population will inevitably rise above zero. It is the ultimate genetic destiny of mankind and is not a dream owned by any given nation. If we do not step up and accept our destiny, then someone else will – and soon.

Reason Number 3

Science and research have an obligation to lead the way in this important and inevitable endeavor. The delicate ecosystems beneath the oceans must be evaluated with extreme vigilance as we enter this watery world, and man's impact must be appraised with care as more and more interests set their sights on this new frontier. The first steps will be the most important steps and they should be accomplished with all eyes focused on the protection of the marine environment by dedicated scientists.

Reason Number 4

Unlike the land areas which are fixed, the marine environment is uniquely mobile. When located properly in a moving current, a single, fixed, manned undersea station can evaluate thousands of miles of ocean annually as it flows by, literally orbiting overhead.

One manned undersea station, if properly placed, could monitor half a hemisphere of the world's oceans from a single vantage point.

Reason Number 5
Humankind is attracted to the beautiful and exotic places of our solar system. Just as men will be attracted to one day live on the cliff sides of the Valley of the Mariner's on Mars, mankind will be attracted to live in the beauty of the undersea regions of our own planet.

Reason Number 6
The undersea regions of the world will one day supply more of the world's resources than are currently supplied by the regions ashore. How can this be? Because more than 70% of all the earth's resources are located there – and most of them remain undiscovered. Learning to live undersea permanently will be a requirement not just to recover these resources, but also to protect and monitor the areas that will one day be opened up to mining and exploitation.

Reason Number 7
New nations and empires beneath the sea are inevitable. That fact is just as probable as was the vision of new nations opening on the North American continents to the Europeans in the 15th century. As unlikely and laughable as it sounded then, the world's political, strategic and military power ultimately shifted there. There is no reason to believe that this will not also happen when the oceans are opened and its vast resources are ultimately utilized. The questions that nations were forced to ask in the past are still relevant today. Do we join in this process or do we recede into historic insignificance, fueled by arrogance and misplaced imaginings of superiority?

Reason Number 8
There are many reasons to shrink back from this enterprise: fear of misstep; fear of a politically–incorrect misinterpretation; fear of damaging the environment; and fear of failure. All of these fears are based on past national and international experiences that have proven that these are all possible consequences of poorly planned or executed actions, or just fate. But in reality, these fears accompany all endeavors of any kind. The greater the reward, the

greater is the specter of failure. When considering failure, I remember what Thomas Edison said: "I have not failed. I've just found 10,000 ways that won't work." In the end, national and organizational courage either overcomes timidity or is ruled by it. Nations must overcome their fear and triumph, or they will die and be replaced by those peoples who have the courage to step out and conquer new frontiers against all odds and resistance.

So, there you have it, my dear reader. Everybody's list is right here to analyze and to consider. But remember one most important thing – just like my mother told me as I was growing up – in the end, "People always wind up doing the things they *want* to do." And that is all the reason they ever need and should need in a free society. Mom and Sir Edmund had a lot of common wisdom that fit perfectly with their uncluttered grip of common sense.

AQUANAUTS AND AQUATICANS

"If a man does not keep pace with his companions, perhaps it is because he hears a different drummer. Let him step to the music he hears, however measured or far away."
Henry David Thoreau

S ince the first 24 hour stay under the sea, there have been more than 7,000 people who have earned their aquanaut certifications in some 69 known habitats by living and working undersea in those habitats for more than 24 hours. These numbers are rough estimates – since careful records were not kept – but this data is close.

As of this writing, of the 69 habitats that have been built, the majority of aquanauts have been certified in only three of them. The *Jules' Undersea Lodge* holds the record of more than 3,800 recreational aquanaut certifications. The *MarineLab* habitat, located near it, has certified more than 2,500 science and research aquanauts. The *Hydrolab* 'housed' over 500 aquanauts from 1966-

1984, while the remaining 67 habitats certified the other estimated 200 plus aquanauts. Again, these numbers are estimates. Also note that the *Hydrolab* numbers do not indicate how many aquanauts returned for repeat missions.

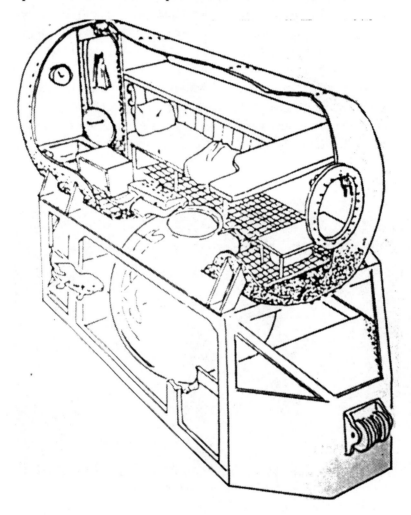

MARINELAB UNDERSEA HABITAT

Of these certifications, the vast majority were for single 24 hour periods. Less than five percent of aquanauts have stayed beneath the surface for periods greater than a single day and less than one percent for up to a week. Records indicate than fewer than 25 individuals have lived and worked in a habitat for up to 30 days and there are probably fewer than 15 individuals who have accumulated more than 30 days living and working beneath the sea in a habitat, per our estimates.

As I stated before, Chris Olstad has logged more time than any other known individual as an aquanaut. He has worked practically every day in an undersea habitat, having logged thousands of man-days in nearly 25 years as an aquanaut. Yet there is not a single Aquatican and there has never been. An Aquatican, again, is an individual who has moved into the sea as a permanent resident. And of the 7,000 plus aquanauts in history, all of them have ended their missions and gone home on the land at the end of the day. While Chris Olstad is the closest person to a real Aquatican in history, he still drives his blue Blazer to his home on land each evening he is not on an underwater mission.

JULES' UNDERSEA LODGE © MRDF

The difference is not at all trivial.

While .000001 percent of all people alive today are aquanauts, there is not a single permanent resident of the undersea realm called Aquatica and there has never been. The reason for that has been discussed: mankind explored using a misdirected approach and then abandoned the push into Aquatica. The requirement for starting back again is a different approach: protecting the human against environmental extremes.

The remainder of this book will explore the building of Aquatica and the initial movement of the first permanent humans into this new world. The first requirement before any Aquatican can leave the shores of the planet is having a place to go! And that place must provide everything the land dweller has in addition to protection from the remote and extreme environment.

What makes an Aquatican unique from an aquanaut? The Aquatican's permanent residence is undersea. If an Aquatican surfaces do they lose their designation? Of course not! If an American travels to France, does he lose his citizenship? No, he does not. The Aquatican may even work his day job ashore; but if his permanent residence is in an undersea colony, outpost or anywhere in Aquatica, then that individual is an Aquatican. Obviously, the difference is in where the individual lives and where he designates his official permanent address. If it is ashore, then he cannot be an Aquatican.

An Aquatican lives in his fixed undersea home (or underwater home, since the dominion of Aquatica is in any part of the world's watery regions) permanently. He is a citizen of the deep and no longer a resident of the land. That individual's home is called an undersea or underwater habitat. That home gets its air/oxygen from the surface, from the sea, or another advanced life support system and does not depend on an open window or vent to the outside air.

The first Aquaticans will be tied to the land and dependent on land-based resources for a significant period of time, just as the earliest colonies were dependent on re-supply from their home nations. But as the Aquatican settles in and begins to count on *in situ* resources, and as technology advances, his dependence will wane until the Aquatican citizen is entirely capable of harvesting all his requirements from the undersea world.

Soon, the number of Aquaticans will begin to grow from the first citizen to thousands – and beyond.

GOING TO STAY

"They conquer who believe they can."
Goethe

I have been on many undersea aquanaut missions – I have logged exactly 15 of them – for more than 30 days living and working undersea.   They were all wonderful and exciting in their own individual ways. The shortest was just over 24 hours (by definition) and the longest was 11 days and 12 hours.   They all began the same way – with intense planning; sometimes designing and building the equipment that went along with the mission. On a few of them, there were sessions in pools and local lakes preparing for the mission. Then there were periods of training in classroom type environments when members of the crew met to discuss elements of the mission so that we were all on the same page.

After all the training and preparation, there was moving my entire family to the site.   Since my children were home schooled at the time, transporting the family for a six week mission away from home was somewhat involved. It meant uprooting the whole family, including all school materials, packing huge bags and even tossing in the dog, then relocating to the mission site.

Once we arrived on the site, the other team members would begin to appear as well. On several undersea missions, the entire assembled team would consist of as many as a dozen or more members – each with their own set of equipment and baggage. As the missions wore on over the weeks, the team would grow and shrink, depending on the activities that were scheduled from day to day. The day before a mission officially began, the team would travel to the grocery store and purchase our frozen meals for the period of time we would be underwater. They would then be delivered to us each day by the duty aquanaut.

At the appointed hour, I would lower myself under the water, swim to the habitat and begin the mission. At the end of the mission, I would halt at 15 feet for a half hour decompression safety stop, then return to the surface for debriefing. After the debriefing, we would slowly pack our gear and return home. At the completion of the appointed time, the team would be gone, returning to their lives and the professions they had only temporarily left.

Each mission began and ended just like that. Many weeks before we departed from home, we knew the drill. And as team leader for each one of them, I knew the mission inside and out. I even knew how much the mission cost in dollars per hour – with equipment costs and without. But as a manager, I also knew that the flow of time, money and people had a starting point and an ending point. Each mission was planned for, executed and finished on a schedule and on a budget.

The point was – there was a mission start and eventually there was a time when the clock ticked down to zero. At that moment, we were always forced to leave our undersea world by the sheer realities of life and the world we live in. That same story has repeated itself in history more than 7,000 times. Each mission has a start time and an end time. Each mission ends with the human returning to the dry land above.

But one day, it will not be that way.

One day in the very near future, the aquanaut will make his preparations in a whole new way. One day, the aquanaut and his team will sell their homes ashore. They will cast off the dock, motor out to their undersea home and enter the water as citizens of the deep. These aquanauts will on that day become the first Aquaticans. They will cast off with the intent of living permanently undersea.

They will dive down to a class of habitats that have never been built before – a class of undersea habitats just now being designed. They are undersea habitats designed for permanence, not for individual missions that start and end on a schedule. These new habitats will be specifically designed for constant use, just like a home on land. And some of them will be designed for families.

In the next few years, many things will radically change in the aquanaut community. Aquanauts of a new breed will enter the picture. They will be aquanauts who are thinking permanence – dreaming and even planning to become the first Aquatican citizens in history. They will dare to imagine themselves as the first citizens of a whole new undersea empire yet unborn.

These 21st century aquanauts will be empowered to do so by one primary change in the status quo. Up until now, the terms have not been in place for people of like mind to even begin the discussion. Until now, aquanauts have been stuck at the dead end of a misdirected path of exploration with no place to philosophically turn. Until this very moment, there have been no designers of permanent habitats, no plans to build and launch them and no colonies in which to settle. But with the publication of these words, all that has changed. The blind alleyway has been eliminated. The way is now cleared.

There is sure to be some complaining by a few of the old school aquanauts who will never believe there ever was a misdirected path in the first place. But that is to be expected. The old school has always clashed with new ideas. Said Albert Einstein of this reality: "Great spirits have always encountered violent opposition from mediocre minds." And just who are the 'great spirits'? Anyone who, like Albert Einstein, dares to think outside traditional boxes and paint whole new realities and whole new worlds with their minds.

When the new generation of aquanauts dare to begin thinking and speaking and acting like Aquaticans, there is definitely going to be a stir. It is a subject with built in controversy. For example, I was recently interviewed by an Australian Broadcasting Company News reporter about Lloyd Godson's unique 2007 underwater mission. As soon as I mentioned a permanent human presence undersea, the reporter instantly changed the subject and wanted to know more. Of course, her first question was, predictably, "Why would anyone ever want to do that?" as though the whole notion was some verbalized form of insanity. The idea of Aquaticans is going to take a lot of 'getting used to' by a whole lot of people, and some of the old school aquanauts are going to have as many or more problems as the uninitiated.

Going to permanently stay will require that the aquanaut decide to give up everything they know for the complete unknown. While many aquanauts have been under for a single day, and a much smaller percentage for more than that, it is quite another plan to go down under with the intention of staying. And knowing what you are getting into is hardly decided and set in stone by a single 24 hour visit.

It is the designed purpose of this book to continue to build on this amazing idea that has never actually been carried out by any human. What happens when people decide to cut their ties with the surface for good? What kind of decisions will they have to make? What do they bring with them and what will they have to leave behind?

This decision making process, while unique to the Aquatican citizens-to-be, is not at all exclusive in history. When the Europeans discovered there was a new world, in the very first waves they came across by the thousands and many had no thought of returning to the old world again. It is quite true than countless in those first ships were the indigent and the desperate with nothing to lose. It is equally true that many were not. Some of them were adventurers, explorers and gentlemen out to make a name for themselves and find their fortunes. But they all came after having gone through the same process – and the giving up of all they had known for an uncertain dream. The process of populating Aquatica will not be much different and the people who go there to stay will have to make exactly the same choices.

The Aquaticans will leave a world of security and relative certainty for a far different environment. For example, they will leave a place where breathing is completely taken for granted. They will leave an environment where no one ever worries about carbon dioxide or oxygen or the prospect of imminent drowning. They will no longer be able to leave their 'home' with all the ease and thought of just opening a door and walking out. Depending on where their colony is, they may or may not be able to leave their habitat and have anywhere else to go except, perhaps, another undersea habitat in their cluster. And they will move to an environment that has the capacity to change in an instant from clear and peaceful to zero visibility with uncertain currents and poorly understood undersea dynamics.

The Aquatican citizen will at first be highly dependent on deliveries from shore based assets and they may or may not experience periods of rationing as they wait for delayed or even lost shipments.

And yet, the Aquatican will also experience many things land dwellers will never be able to enjoy. They will live in a world that is far different in colors, appearance, 'weather' patterns, lighting, and in most other aspects that are completely alien and unknown to the land dweller. They will be citizens of a new world so radically different that the only other way to match the change is if the human had gone to live on another planet. The undersea kingdom is indeed that different and altogether alien.

In 1991, when the League of the New Worlds Corporation (our 501(c)3 marine research group dedicated to establishing the first permanent undersea colony) was formed and we began to try and communicate this aspect of the new world of permanent ocean dwellers, I endeavored to describe the unique nature of the quest:

"For the first time in history, mankind will permanently depart the shores of his planet for an alien environment. At no time has man ever departed the safety and security of the dry lands or the comforting blanket of air for a world so alien and so different that he will be required to provide the very air that he breathes."[a]

That is what the Aquatican decides to give up – and to embrace as a permanent lifestyle: the known for the unknown and security for a measurable degree of uncertainty.

One day, the first Aquaticans will drive to the pier and unload their gear, leaving all they have ever known behind them forever. They will toss their belongings into the boat and will step off the pier as citizens of the land for the last time. The vessel will speed them to the waters over their habitat. They may well look over their shoulders at the world they have chosen to leave behind. Perhaps they will cluster together for a humble prayer. Then they will suit up, drop into the liquid deep, and disappear into their world – the first citizens of the new empire of Aquatica.

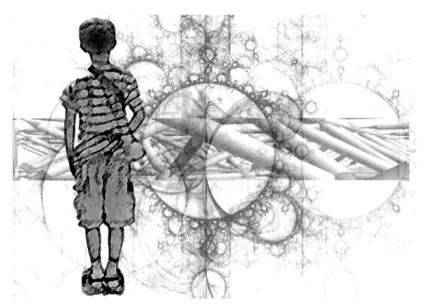

THE UNDERSEA ENVIRONMENT

"When winds are raging o'er the upper ocean and billows wild contend with angry roar, 'Tis said, far down beneath the wild commotion, that peaceful stillness reigneth evermore.
Far, far beneath, the noise of tempests dieth and silver waves chime ever peacefully,
And no rude storm, how fierce soe'er it flyeth, disturbs the Sabbath of that deeper sea."
Harriet Beecher Stowe

 H ere is an interesting comment written on a recent Internet blog about the whole idea of living undersea:

"...I wonder whether I'd actually like to live permanently in a world where I never saw a sunset or sunrise, and never felt the wind on my face?"[a]

You cannot help but focus on the radical differences between the environment we have all lived in each day of our lives and another, quite different and even alien place. It is quite interesting that I have never heard this identical comment about living in space, or on the Moon where there is only a single sunrise and sunset per month and no wind. The fact is, the ocean is a very different environment – radically different than on land.

The writer of the above comment spent a few moments and contemplated what it would be like to live in such a place, and very objectively described the first things he would miss – sunrises, sunsets and the feel of moving air on his skin. His comments were honest and describe not only what would be missing, but also the emotional links that every human has with their environment.

We should focus on the obvious difference at once: the environment of Aquatica is filled with a liquid – water – that excludes the gas mixture we call air. It is this pervasive blanket of water that causes the environment to be different, even alien, from all that we know. But water does not make the environment evil. It is not overtly sinister and it can be completely understood.

However, in any environment where air is excluded, either the undersea environment or the space environment, the human faces immediate challenges. But just as those challenges are manageable in space, they are also equally manageable undersea. While it is true that many people die each and every year by drowning, multiplied millions more die by many other causes without even being near the water. Water and imminent drowning are not always invariably linked. Millions of swimmers, surfers, boaters, skiers and SCUBA divers go into that environment and come home safely each day.[7] Therefore, the undersea environment is alien – but not at all an impossible and impenetrable barrier to human activities or even long term colonization.

In physical terms, water is obviously a fluid medium. But less obvious, physically speaking, air is also a fluid medium. It is this special understanding of the fluid properties of air that allowed the science of aerodynamics to be developed.

---

[7] In my first SCUBA class my instructor taught me his "Number One Rule of Diving: Be sure and plan as many exits as you plan entries."

Air and water have many things in common: The more air the human lives under, the higher the pressure. The air pressure on a human at sea level is 14.7 pounds per square inch – also known as a unit of one atmosphere of pressure (ATM). This pressure is caused by the weight of the entire atmosphere pressing on the human form. If my body has a surface area of 3,270 square inches, then the combined air pressure on my body is 48,069 pounds of force! Fortunately, my lungs are open and every cavity of my body is pressurized internally to the same pressure, so I do not 'notice' this pressure, even though it is present and acting on me 24 hours a day.

Since water is much denser than air – about 784 times more so – the aquanaut only has to submerge himself to 33 feet to equal the same pressure as 100 miles of our atmosphere from the ground to space!

The surface dweller and the aquanaut both have to deal with pressure in exactly the same way. As the aquanaut breathes compressed air at the same pressure of his surrounding, he therefore does not notice any more pressure difference than does the surface dweller – even if the pressure on his body is two, three or four times that of the shore based human. An aquanaut outside at 132 feet beneath the surface breathing on SCUBA gear will have almost 200,000 pounds of force pressing against their body, but will not notice it at all! And yet, in his environment, the pressure changes much faster than on the surface because of the density of the medium in which he lives. But just as land dwelling mountaineers have discovered, radical changes in pressure have rather dramatic effects on human physiology. This is discussed in greater depth in the next chapter.

Perhaps one of the most immediate changes noticeable to the human when he enters the undersea world is light and color. On several occasions I have cut myself as an aquanaut. At around 30 feet, most of the reds have already disappeared. So as I hold my hand or finger up and watch the thin trail of blood curl in the water, it is somewhat disconcerting to see that at even that shallow depth, my blood has turned to a dull green! This happens because water absorbs the wavelengths of light that cause us to see color.

The reds go first, since water absorbs red light – the longest wavelength - first.

As light passes through the water, some of the energy in light is lost in the process. That specific energy is sometimes called the 'spectral energy' or the energy of individual color. This causes colors to appear lost as objects are viewed at deeper depths or from further away in the water. Further, the wavelengths that make up our perception of colors are also absorbed differently. The 'length' of the light wave determines how fast the color is absorbed. Red has the longest wavelength, more than 700 nm (one 'nm' represents one nanometer, which is one millionth of a meter). Following red is orange, somewhere in-between 700nm and 600nm. Then yellow and so on, all the way through the spectral ROYGBIV to the shorter wavelengths of the blues and purples at around 400 nm. Here is a table that represents how colors change as depths change:

Color visibility at changing depths:
Reds are visible from 0' - 30'
Oranges are visible from 0' - 45'
Yellows are visible from 0' - 60'
Greens are visible from 0'- 80'
Blues are visible from 0' - 100'
Purples are visible from 0' - 120'
From 120' and deeper, the undersea world loses all
its color and fades to grays and black.

Color visibility is also dependent on other things – such as particulates in the water that are also absorbing light – so that on some days colors appear differently than on other days, depending on the overall water condition.

As you go deeper into the water, light also decreases. The eye is a wonderfully compensating mechanism, and, as you enter the water and descend, your eye balances for the dimming light so that the aquanaut will not necessarily notice the dramatic loss of light, even in clear water. Technically speaking, when light penetrates water its intensity decreases exponentially with increasing depth. This process is known as attenuation.

Because the process is described as non-linear or exponential, the light and color fades dramatically at shallow depths, and then it tapers off much more slowly to complete darkness as all the light is absorbed by the water. Here is a table that demonstrates that effect in perfectly clear ocean water at high noon:

- Beyond 120 feet all color is lost.
- At 0 to 300 feet the water and surroundings are light blue to gray.
- From 300 to 600 feet objects fade to dark blue indicating the lowest limits of visible light penetration.
- From 600 feet to 3000 feet the ocean and objects are almost totally black indicating a perpetual 'twilight zone'.
- Beyond 3000 feet, there is no light penetration at all.[b]

Hence, permanent residents of Aquatica will adjust to a world of dimmer light and an exterior world that is illuminated at the shallower depths by predominantly greens and blues. Aquaticans who live in deeper colonies will enter a world much like space dwellers, with no light except for the passing bioluminescent life and the light they create for themselves.

Undersea dwellers also face an effect well known to their surface dwelling friends. Some days are clear and visibility extends for a long distance under the bright sunshine. But on other days, when the sun does not shine on the surface, the undersea light loss is quite dramatic.

As I served as an aquanaut near 30 feet, I noticed the loss of sunlight on cloudy days to be dramatic. My eyes easily adjusted to the diminished light on a clear day at 30 feet so that my surroundings appeared relatively 'normal'. But on cloudy days, the loss of light was far more dramatic. On these days, almost all color vanished and the world outside became one dominated by grays.

On the surface, the eye adjusts to the blocking of sunlight by clouds so that a dreary day is only a dreary day and not necessarily 'dark', depending on the thickness of the overhead cloud layers. But undersea, it is a wholly different story. The dual effect of clouds and water blocking the sunlight results in a dramatic impact on the eye whose limit of compensation is truly challenged in these conditions. Hence, a cloudy day above the surface is always a dramatically dreary day undersea.

But clouds are not the only light blocking mechanism undersea. Visibility, light and color are also noticeably affected by water conditions and water quality. Even on a brightly lit sunny day on the surface, water conditions undersea can cause the visibility to drop – sometimes even to zero. It is an effect that the surface dweller would call 'fog'. Undersea, these conditions effect our 'viz', which is short for 'visibility'.

The water visibility is determined by a wide number of factors, not all of which are easy to understand from a fixed position in the ocean. Like a land bound observer, sometimes the fog rolls in and sometimes it just 'appears' as the conditions change. The same thing is true underwater, but for many widely different reasons.

Sometimes visibility is reduced by cloud cover overhead, as was discussed previously. If the habitat is located near shore, often rain and the run off caused by it will cloud the water and reduce visibility. In relatively shallow estuaries, wind can drive currents which will scour the bottom. Even in deeper areas, if the estuaries are surrounded by shallows, the silt will easily move by wind driven waves and encompass the entire body of water.

If you look at satellite or high altitude photographs, it is easy to see how areas of the ocean near land are affected by runoff that carries silt and other land depositions with it into the water. These waters run into the sea and interact with the currents and even the Coriolis Force (the force exerted on a body when it moves in a rotating reference frame) that tends to mix the discolored, slit laden water in huge circulating gyres or streams with the clear ocean water.

Aside from silt and runoff, visibility can be reduced even in the open ocean by blooms of plants in a given area. In parts of the North Atlantic Ocean, great mats of sea grasses, such as the sargassum weed, can sometimes float over a given area and sharply reduce the visibility for days or even weeks. And in the

North Pacific, gigantic mats of kelp can also float over expansive regions.

In open ocean areas, the effects of visibility reduction are fairly uncommon. However, settlements that are placed nearer the shorelines will, generally speaking, find their viz reduced by shore based influences on a more regular basis.

Since 1865, undersea visibility has been measured by a device called a Secchi disk (pronounced 'sekie'). The Secchi disk is very simple to use and to interpret. It is a disk eight inches in diameter divided into four quadrants painted black and white on opposite quadrants. From the surface, the disk is lowered into the water until it disappears. The distance reading on the line determines the visibility in feet. Underwater, one diver holds the end of a line while another diver swims slowly away. When the disk disappears from view, that is the visibility in feet at that depth. For standardized results, these readings are usually taken at a high sun angle between 10 AM and 2 PM.

Since the undersea environment for the human settler is far different than the land environment, undersea visibility comes with many unusual effects. Because the aquanaut is not limited by a two dimensional surface, if conditions are right he may simply float above a layer of reduced visibility and work above the layer in clear water. On one research dive in the open ocean, our target lay at 65 feet below the surface. But around 40 feet, the water was obscured by a roiling gray cloud of floating haze. On the bottom, the visibility was reduced to nearly zero. So I chose to hover above the murk and handle equipment where I could see what I was doing while another diver remained on the bottom deploying the equipment and a third shuttled equipment back and forth from me and the diver below. If the visibility is fortuitously reduced in layers, aquanauts can chose to work above or below the layers, since they are not at all confined by two dimensions.

On land, there are gaseous currents in the air called wind. Unless the wind is very strong – typically over 50 knots (57.6 miles per hour) – humans can at least walk and function in its stream. However, underwater, the movement of the aqueous fluid (water) is called the 'current'. Any current over 3-4 knots (3.4 - 4.6 miles per hour) is beyond the human capacity to move against except for very short periods and is considered by most experienced divers to be dangerous.

In the undersea environment, there are permanent currents – such as the Gulf Stream, the North Pacific Drift and Humboldt, just to name a few. In these areas, the current is relatively constant and typically unchanging. The Gulf Stream holds the record for the fastest ocean current in the world, reaching an average velocity of two to four knots, although speeds of eight knots have been reported on rare occasions. The Gulf Stream is characteristic of other broad ocean currents, and its greatest velocities are confined to the surface waters down to about 600 feet with current velocity declining with depth.

However, in other places, there are currents caused by tidal flow, and these are by far the strongest currents in the ocean. Tidal currents can be very strong – such as the Saltstraumen Sound in Norway whose tidal current runs up to 22 knots! Obviously, these are limited to inland and near shore areas where the currents are influenced by inlets and other land based features.

In nearly any location in the ocean, there are currents to deal with. In areas that typically have minimal currents, there are occasions when these regions are influenced by larger nearby currents that sweep in hemispheric sized arcs called 'gyres'. Sometimes these gyres change and move due to many influences and sweep over areas normally not influenced by currents.

Currents will definitely always influence habitat design and placement. Further, locating in or near a current is not necessarily a bad idea. Power can be harvested from a constant current, as will be discussed in detail later. Current generally ensures good visibility and can help generate large amounts of data about the entire hemisphere over which it circulates, so that a well placed undersea scientific station can perhaps harvest a quarter of the planet's oceanographic constants while patiently sitting still as the ocean literally orbits overhead! Such a concept was essential to the placement design for the League of the New World's Challenger Station – the world's first permanent undersea colony design. According to our proposal, that colony will be placed in the Gulf Stream to maximize scientific return.

The air-sea interface can be a violent place. Even on a typical day, the air-sea interface can be no fun. With bouncing waves, wind and spray, the air-water interface can be difficult to contend with.

When I was a United States Naval Officer aboard ships that ranged the Pacific basin from the Bearing Sea to China to the equator, I never tired of the many faces of the surface of the sea. On some days, in the vast expanse of ocean thousands of miles from the nearest land, the ocean could be mirror flat with no discernable wind. Sometimes, just an hour later, our ship would be tossed about in gale force winds and twenty foot seas. Each day it seemed the sea's face changed, even though from horizon to horizon it was still just a vast, open plane of water.

But just as the forces of wind and waves affect the surface of the sea, these energies are also transferred below. And just how this energy is transferred is of vital importance to the dweller beneath the surface.

On the surface of the ocean there are waves. Not every wave is the same, yet all waves have the exact same physical properties. It is how these physical properties change that determine the wave's characteristics and that sometimes dramatically affects what happens below the surface as the wave passes overhead.

Every wave has a wave height and a wave length. These can be very different from wave to wave.

If you were living in a sphere at the surface of the sea, as each wave passed, the sphere would appear to move up and down in a harmonious, sinusoidal based circle defined by the wave height and length. But if you began to move your sphere down deeper into the water, you would find the sphere's motion begin to diminish with depth. The deeper the sphere, the smaller the circles. At a given depth, eventually the sphere would stop moving altogether as the energy of the wave was entirely spent above it.

The sphere stopping at the base of the wave is defined as exactly the wave's length. Very short wavelength waves are called *chop*. Long wavelength waves are called *swell*. *Fetch* is a term that defines not only the wave's length but also its orientation. One *wave period* is the time it takes for one wave crest to pass to another wave crest. Wave periods are determined by the fetch, wind speed and duration.

Now having all that information, imagine that you are an aquanaut living in a bubble. It would be best for you, and especially for your habitat, if you did not move about at all – hence, the deeper you place your habitat, the less energy the passing waves exert on you and the structure you are living in. On an average day, it is probably safe to place your habitat at 50 to 60 feet or so. At that depth, surface action is so minimized that it is probably hardly noticed. But if there are long period swells passing overhead – as is common on some coastlines – then the wave energy is going to affect your habitat at much deeper depths. For example, if a long period swell passes overhead with a wave length of 300 feet (not uncommon), then that wave can be felt at a depth of 150 feet!

The longer the swell period is an indication that the wind has transferred more energy into that wave. Long period swells (generally defined as waves with a period of more than 12-14 seconds) are able to conserve their energy as they travel extended distances across the sea. Short-period swells are those who have wave periods of less than 12-14 seconds. They are generally steeper as they travel across the ocean and, therefore, decay more easily due to opposing winds and seas. However, long period swells travel with more energy – including below the surface – and are not as steep so they can more easily pass through areas of opposing winds with very little affect.

The depth at which the waves actually touch the ocean floor is one-half the wavelength between wave crests. Wavelength and swell period are related, so you can use the swell period to calculate the exact depth at which the waves will begin to touch the seafloor. In order to calculate this, square the number of seconds between swells and then multiply by 2.56. This will tell you the depth the energy begins to touch the bottom.

For instance, a 22 second swell will be felt on the ocean floor at 1,239 feet. The energy from a 16 second wave will touch bottom at 655 feet. As you can see, the shorter the wave period, the less effect of its depth. A 12 second wave will express its energy down to 367 feet; a 10 second wave down to 256 feet; a seven second wave at 164 feet; a six second wave at 92 feet and so on. As noted above, longer period swells affect the ocean floor much more than short-period swells. For that reason, long-period swells are sometimes referred to as ground swells.[c]

In ambient pressure habitats or hybrid habitats with moonpools, long period swells may cause the water level in the habitat's moonpool to cycle up and down on each swell, depending on depth and wavelength. This will cause the ambient habitat's pressure to rise and fall on each cycle which is quite uncomfortable to the aquanauts.

Yet, if the long period swell was even more dramatic due to a longer wavelength, the level in the moon pool could rise dramatically, possibly flooding the dry areas of the habitat on each surge, damaging or even washing away equipment depending completely on the characteristic of the wave. Further, on each rise and fall of the wave, the habitat itself innately attempts to rise and fall with it, just like the aquanaut in the sphere. Hence, the habitat must be very negatively weighted or the wave action will actually lift it and drop it back to the seafloor on each and every pass! A very heavily ballasted habitat is strongly recommended to prevent this effect from damaging or destroying it.

For example, The Soviet habitat *Chernomor* was so lightly ballasted that the habitat was picked up off the bottom of the Black Sea at 50 feet and was actually washed on to the shore during a storm! Fortunately, its crew was locked in under pressure and the hull was not breached during the incident. They opened their hatch, climbed up the beach, were rushed to a decompression chamber and all survived – incredibly seasick but safe.

Further, I recommend that every moonpool area in the open sea have the capacity of being sealed off from the outside to prevent flooding or dangerous hydraulically generated air slugs from damaging the habitat, not to mention the vestibular hell generated on the aquanauts by the effect.

CHERNOMOR HABITAT

Knowing the swell data enables us to calculate the exact effect of those swells in the moonpool of our habitat. A wave's energy is proportional to the square of its height (potential). Thus a 4 foot high wave has 16 times more energy than a 1 foot high wave. Likewise, as you look below the surface, a wave's energy exponentially decreases as you move deeper under water. Hence, a wave with a height of 10 feet and a period of 10 seconds will cause an oscillating rise and fall of 5 feet at 128 feet depth, but only 2.5 feet at 192 feet depth , 1.25 feet at 224 feet depth, around half a foot at 243 feet and so forth. At 256 feet, it is not felt at all.

Another hazard of the long period swell is an effect called 'scouring', where the sand or bottom material is picked up and moved by the energy of the passing undersea swells. This effect can literally undermine the habitat's placement by hydro-dynamically removing the support from under the habitat, causing it to lean off center, exposing the base to swell energy and eventually causing the habitat to become unstable or unfit for occupation.

Long period swells probably pose one of the greatest natural risks in day to day habitat operations. Long period swells are best engineered for by a series of habitat design characteristics.

Certainly, if the habitat is part of an undersea community or if the habitat is placed for permanence, every potential hazard should be designed for, especially the day to day possibility of long period swells.

The best solution is heavily ballasting the habitat and the employment of various anchoring strategies so that no conceivable wave action will ever lift it off the seafloor. The second solution is hydrodynamic design so that the habitat is built to deflect swell energy and not offer surfaces for lifting or 'sail area' for the currents to take advantage of.[8] The third is closely related to the second – to place ballasted rip-rap revetment (bags of ballast materials laid in a closed rim around the base of the habitat) in such a way that it offers no entry point for scouring or lifting forces and it also deflects the energy up and away from the face of the habitat. A fourth solution is a thorough understanding of the undersea environment by placement of long term probes to collect data before the habitat is positioned. In that way, the prevailing currents and swells can be thoroughly understood so that the structures can be most advantageously designed, placed and oriented to meet the environmental challenges.

However, all bets are off when it comes to hurricanes and their powerful effects. In hurricane prone waters, permanent undersea colonies placed less than 300 feet in depth will probably want to consider evacuating in an approaching hurricane. Habitats under 300 feet are probably safe from the hurricane's effects.

There is some detailed knowledge about what actually happens undersea during a hurricane due to the work of Dr. Steve Miller with NOAA and the National Underwater Research Center. NOAA/NURC has placed data buoys in 60 feet of water on the ocean floor as hurricanes Hugo (1989), Georges (1998), and Irene (1999) passed. From the data recorded, we now know many of the effects of a hurricane on the ocean floor environment.

---

[8] In 1972, the South African HUNUC habitat, unmanned during its initial deployment, was turned over and destroyed by currents as the habitat featured a high sail area and insufficient ballast. The undersea currents turned the habitat over, carried her across the seafloor and smashed her against the reefs.

Good data notwithstanding, there are plenty of examples of the result of a hurricane sweeping over the undersea environment. The following illustrations are all regarding sunken ships, except one which involves an undersea habitat. Please note that in the case of sunken ships, there is no thought of engineered stability, no negative ballasting, no consideration of positional hydrodynamics, and no regard for placement, orientation or scour prevention. They are generally large pieces of metal lying where they landed when they sank. But what happened to these structures does offer a clear warning of the powerful ability of a hurricane to disturb the undersea environment.

Example One: The 187 foot long *Cayman Salvage Master* was deliberately sunk in 93 feet of water off Key West in 1985 to serve as an artificial reef. She settled to the bottom on her side and remained that way until 2005 when a powerful undersea surge from Hurricane Katrina flipped her upright.

Example Two: The 9,000 ton, 510 foot *USS Spiegel Grove* was sunk for an artificial reef off the Florida Keys in 2002. Unfortunately, she went down the wrong way and landed on her side to the extreme disappointment of the diving community. However, Hurricane Dennis came by and righted the massive ship, just as the human salvage team had intended! That amazing feat was accomplished by a category four hurricane passing 200 miles away!

Example Three: The 194 foot German freighter *Mercedes* was sunk in 1985 in 97 feet of water as an artificial reef off the southeast Florida coast. Hurricane Andrew nearly tore the ship in half in 1993.

Example Four: The 227 foot *Jim Atria* was sunk as an artificial reef in 1987. When the ship settled to the bottom, she landed on her side in 112 feet of water. In August of 1993, Hurricane Andrew not only righted the vessel but it then pushed her into 132 feet of water!

Example Five: The *Aquarius* undersea habitat administered by NOAA sat in 60 feet of water near Key Largo. In 2005, Hurricane Rita passed south of the Keys before entering the Gulf of Mexico. Although only a category one hurricane at the time, wave heights reached nearly 30 feet off Key Largo, over 100 miles away from the storm. The surge and constant wind-driven currents from the east-southeast moved *Aquarius* eight to ten feet and threatened to tip the habitat over.

What is the precise physical mechanism that wrecks all this havoc on the ocean floor? A hurricane does not generate magical or evil forces. What it does is magnify what are normally powerful mechanisms into extraordinarily powerful mechanisms.   For example, a hurricane generates mid to long period swells (generally a 9-13 second period) over relatively short ocean distances. But their wave height – which determines the wave's energy – can be enormous. Hence, in even a relatively minimal long period wave with a massive wave height, the energy transferred to deep water packed within a longer period swell can have an extraordinary effect beneath the surface.

Yet, the energy in a wave is in its height so, that in storms that generate extraordinary waves, the effect under the surface is magnified. In hurricane Ivan (2004) waves were reported with heights reaching 54 feet and maximum crest-to-trough individual wave heights of 91 feet. Analysis suggests that significant wave heights likely surpassed 69 feet and that maximum crest-to-trough individual wave heights actually exceeded 132 feet near the eyewall![d]

Even in these extraordinary events, the effect is known and simply magnified, therefore it can be understood. And if it can be understood then it can be planned for to prevent the wholesale destruction of undersea communities built in hurricane prone areas.

The wave height (energy) and wave length (expressed depth of the energy) work together to form a cyclically moving mass of water that impinges on any structure in a defined cyclic pattern. This mass of water will impose itself on a static structure and act upon it with a hydrodynamic force equal to the energy (wave height) and vectored geometrically by the wave period. In other words, if there is a sunken ship lying on the bottom whose flat deck is exposed to the geometrical force of the wave, the power expressed by the moving mass of water will try and push, lift and twist the wreck according to the various geometries exposed to the cyclic energy.

It is interesting to note, however, that even after massive hurricanes have passed over broad reef areas, some large reef structures appear to escape unscathed. Brain corals and other rounded shapes seem to weather hurricane forces well. The reason seems to be obvious: their shape allows the energy to pass around and over them from any given orientation and their anchor is tight. They have few flat planes and nearly no cusps to capture the force of the passing energy. The sunken ship, however, is built such that only its hull is hydrodynamic, which may explain why the *Jim Atria*, the *Cayman Salvage Master* and the 9,000 ton, 510 foot *USS Spiegel Grove* were all up-righted in the undersea tempest rather than torn apart. Had they lain in another angle to the surge, their fates might have been different, such as in the case of the German freighter *Mercedes*.

Next generation habitat designs will include all that is needed to weather the toughest storms, featuring: surfaces that deflect surge rather than capture it; shapes that pass hydrodynamic forces rather than resist them; super secure anchorage; and protection against scouring.

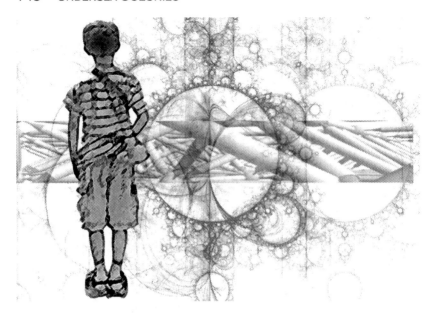

OUTSIDE AND INSIDE

"Knowledge of the oceans is more than a matter of curiosity.
Our very survival may hinge upon it."
President John F. Kennedy

I will never forget my first visit to an undersea habitat nor will I forget the ideas and feelings swirling about in my mind. I remember pulling myself up into the large wetroom and sitting on the side of the moonpool, dripping wet. I recall the smell and the lights and the sound. My ears had not completely cleared, so the sound was just a bit muffled. The lights lit the space brilliantly and the color of the green water beneath the opening below my feet was beautiful.

I stood up with seawater streaming off my wetsuit, slid out of my gear, took a warm shower, changed into dry clothing and then returned to the wet room. I clearly remember my first thought as I looked down into the opening. As I stood there in dry clothing looking at the opening into the sea, I realized there was no 'easy way' out of this habitat, even though the moonpool was large and open wide.

The bottom beneath me was 30 feet from the surface and to get anywhere required suiting back up, diving into the water and swimming safely to wherever it was that I needed to go next. It was completely unlike the decades of life I had lived onshore where, whenever I wanted to go somewhere – even into the back yard – all I had to do was open the door and walk out. Now, I no longer had that freedom. I noticed it right away.

In terms of restrictions, it was exceptionally minor. We were right at the threshold of decompression depth, so making it back to the surface in the event of an emergency was a very simple matter of jumping into the water and swimming back up with or without equipment. As long as we exhaled all the way back up, we would be fine. It would have taken less than a minute to make it to safety. I have been known to experience some level of claustrophobia, so taking that mental note was probably not one many others would have even noticed. And yet it was very real – my absolute freedom was now restricted and I was paying careful attention to that fact.

At that depth, the restriction is about as minor as it can possibly be. The only constraint to going anywhere – either back to the surface or anywhere underwater – was essentially getting wet. If we were but a few feet deeper, the story would have been significantly different. At a greater depth, we would have been saturated with nitrogen to the point that we would have had to go through a long, complex process of decompression before we could have safely returned to the surface. After saturation below a hatch depth of 21 feet, a direct return to the surface could be permanently crippling or even deadly. Those facts would have added not only to the absolute complication of our freedom, but to the complexity of the psychological condition associated with the whole experience.

In the case of the Aquatican, this restriction of movement is no longer just a brief 'experience' but it becomes a chosen lifestyle, which ramps up yet another notch the need for specific mental conditioning. Over many undersea missions, I began to notice that, in my own mind, there was a process I came to call 'latent stress'. That term is not one that is, as far as I know, an official medical term, but one that I applied to my own personal psychological profile. I realized early on that in order to properly do my job as Mission Commander and Principal Investigator, I at least had to understand myself before I could be responsible for the lives and well being of others.

I defined latent stress as the combined effect of many small stressors building up until I was sensing it all as an undefined, background anxiety and a vague, consistent feeling of nagging discomfort. These stresses included such relative trivia as the conciliation of immediate freedom, but also the safety of many crewmembers for which I was held directly responsible; the success of all our engineering systems which I had personally designed and then convinced my superiors would work as advertised; and the reality that just about anything could, and usually would, go wrong without notice. It amounted to many moments of discomfort as I added them all up. But when I sat quietly and thought about it, I realized the discomfort I felt was actually a combined sensation of many individual elements that had morphed into a single sense of uneasiness. Later, I learned to write the different elements down on paper along with their solutions and the latent stress usually disappeared.

I soon recognized that one of those early elements of latent stress was caused by the nagging feeling that my absolute freedom had been compromised. It was probably aggravated by the concurrent reality of latent claustrophobia, but it was real and tangible and I remember its psychological foundation clearly. I also clearly recall that after a few days undersea, the feeling always completely disappeared as I adjusted to my new surroundings and to the unique freedoms that are not available on land, such as the ability to float in three dimensions in the water.

As you consider the range of Aquatican situations, the freedom to come and go from inside to outside the habitat becomes even more restrictive. For example, even with an open moonpool in an ambient habitat, if the water is cold, the aquanaut will have to dress in a wetsuit or dry suit before venturing out. And in the case of a hybrid habitat with a one atmosphere living section, the aquanaut will have to pass though a lock-out chamber to get to the moonpool. Further, his movement will have to be monitored minute by minute if he expects to return without having to decompress. And in the case of very deep habitats, no one will ever get wet (except in the shower), always venturing outside in hardsuits.

Yet, even the most restrictive scenario is not at all different than for any astronaut on a manned space mission who has equally rigid restrictions on movement from inside to outside. As an Aquatican, just as with any astronaut, this restriction on freedom of movement will necessarily have to be psychologically assimilated. I am assuming that some humans who have lived many years on dry land will never be able to adjust to this, while Aquaticans who have been born into that environment will not know any difference. Further, some severe claustrophobics will never be able to adjust and will probably get the willies just reading these words!

Access to the outside may also be limited by system designs and resource limitations. Independent habitat systems that generate their own atmosphere from whatever mechanism will begin to count cubic feet of breathing gasses as having a tangible value. While surface dwellers have a virtually unlimited supply of air, undersea inhabitants do not! Even if it is compressed into gas bottles and delivered to the site, it will still cost resources. In indefinite missions, this essential cost will be minimized as much as possible, and ways will immediately be designed into operations to stem the losses that occur through leakage and unnecessary use.

For instance, SCUBA gear is one example of a gross and unnecessary waste of gasses. A liter or so is breathed into the aquanaut's lungs on each breath, exhaled out into the water and forever lost. With rebreathing systems however, breathing gas is scrubbed of carbon dioxide, replenished with oxygen and used over and over again. It would be difficult to imagine a permanent colony whose aquanauts relied on SCUBA equipment on any routine basis. If so, access to the outside would be severely restricted since it would be responsible for so much waste of breathing gasses.

Another consideration is habitats which use an air lock-out to pass from a one atmosphere part of the habitat into an ambient pressure section of the habitat. On the way from the depressurized module into the pressurized compartment, compressed air must be introduced into the lock-out chamber before passage into the pressurized compartment. Nothing here is lost. But on the return trip through the same airlock, the compressed air must be reduced to one atmosphere before the aquanaut can return.

The easiest way to do this is to run a rigid line from the airlock to the surface and to dump the pressure to the surface. However, that wastes a volume of air equal to the airlock space less the volume of the aquanaut. In the case of a medium sized airlock at 5 atmospheres (165 feet), that could cost the community around 375 cubic feet of air per passage. Or, on an average busy day, as much as 3,750 cubic feet of air per day or more! That, of course, is a totally unacceptable hemorrhage of air.

To eliminate this loss, next generation habitats will employ powerful air pumps that will pump the air from ambient pressure to cabin pressure back into a storage tank instead of simply releasing it to the surface. This method will cost in terms of complexity of machinery and energy, but precious air will be conserved.

After every consideration is taken into account, aquanauts will settle permanently in Aquatica only when these concepts are carefully understood and engineered. Out of these ideas, the most important design process of all will evolve – the development of a reliable, robust, forgiving and automated life support system. It is this system that will ultimately allow humans to live in such an alien place.

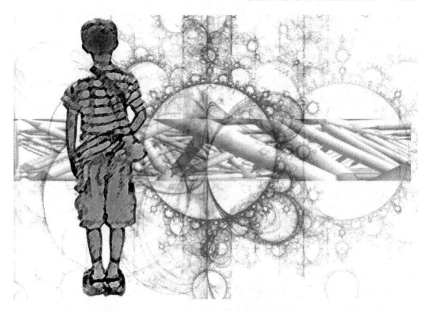

LIFE SUPPORT SYSTEMS AND ALIEN ATMOSPHERES

"OCEAN, n.: A body of water occupying about two-thirds of a world made for man – who has no gills."
Ambrose Bierce 1842

When I was growing up on the Oklahoma plains, I read two books that made a profound difference in my life. One was titled, *The Treasure of the Coral Reef* by Don Stanford and the other, *Pearl Lagoon* by Charles Nordhoff. They were small books written as juvenile fiction. But they painted powerful images of being able to dive undersea that, even as a young teenager lying on my couch in the middle of the North American continent, far from any ocean, still made an indelible impression.

The books were written not long after Jacques Cousteau and Emile Gagnan began to market the SCUBA apparatus they had co-invented. In the stories, they called SCUBA diving, 'lung diving' – probably because the first units were trademarked 'aqua lung'. I am happy 'lung diving' did not permanently affix itself to the sport.

I soon wrote for catalogues from the major manufactures of SCUBA equipment and sighed wistfully at gear that I desperately yearned for. But with an allowance of 50 cents per week, the gear was way over my budget, as well as my father's. Besides, I did not even know how to swim – quite a handicap for someone secretly wanting to build an undersea empire.

One day, after I had worked in my first real job, I saved enough money for my first piece of SCUBA gear – a US Divers mask. Later, I saved enough for a backpack to hold the tank I did not yet have. Eventually, my mother saw I was deadly serious about the whole thing and, unfortunately for her, that it was not going to be a passing fancy. When I showed up one day and modeled for her a weight belt outfitted with lead weights that I had acquired, explaining that I would wear it so I would be sure and sink, she mercifully suggested that I take swimming lessons before this madness continued any further.

My mother, in her infinite patience and knowledge of young teens, decided to enable me to take the first necessary steps to my dream by enrolling me in the Tulsa YMCA swimming program. This, she reasoned aloud, would be the ideal course of action before strapping the lead weights to my belt and jumping into water any more than say – four feet deep. I will never forget the ride to the Y and the ensuing, most humiliating day of my young-teen life.

Since I could not swim a lick, even though I was fourteen, they had no choice but to put me in the beginner Pollywog class. The rest of the kids were about half my age and half my height. Like a school of little piranhas, they ganged up on me and had a very good time at my expense. I left that day with wounded teenage pride and no intention of ever going back. The only problem was, I still could not so much as tread water.

There you have the sad confession. At age 14, I was a YMCA swimming dropout and an abject Pollywog failure.

Shortly thereafter, girls intervened in this dilemma and hormone driven priorities took command of my rational thinking and my carefully chosen path into the abyss of ocean and space.

Soon I found myself an undergraduate student at Oklahoma State University. The whole hormone issue had not resolved itself then (or now), but I somehow learned how to integrate it with the rest of my thinking.

One Saturday afternoon in 1970, out of sheer boredom, I wandered over to the Colvin Physical Education Center and wound up in my bathing suit staring at the magnificent indoor pool. I was alone and the pool was abandoned save for me and the lifeguard. Unfortunately, the depth of most of the pool was over my head and I was very self conscious about not being able to swim. I looked over at the lifeguard and down at the water; over to the lifeguard and down at the water. Then I slipped into the shallowest end and just stood there for a long time.

I was encouraged since the lifeguard was only a few yards away and I was the only one in the pool, so the chances of me actually drowning were fairly low. I began to experiment with what the YMCA teacher had called 'bobbing', then I hand walked myself around the edge of the pool, clinging to the wall when the depth fell below my ability to stand. I have hoped over all these years that the lifeguard was daydreaming and not paying much attention to me and my Pollywog antics. What I have come to realize was that I was in the process of very effectively teaching myself to swim – making some of it up as I went and remembering what I could from my only class at age 14. It may have appeared silly, but it was amazingly effective.

Eventually, the crucial moment came. I walked over to the diving well – 14 feet deep – and lowered myself into the water. I clung to the wall for a long moment, then let go, treading water. It worked. It was easy. There was no screaming or splashing about. I didn't die. I was just calmly and peacefully treading water. From there the rest was easy; kicking from corner to corner and paddling my way edge-to-edge. Eventually the time came to swim all the way across the pool and ultimately to swimming laps. It took no more than three Saturday sessions. I was now surging with enthusiasm and abundant youthful testosterone and fully ready for both the devil *and* the deep blue sea. Before that spring was out, I had purchased my own SCUBA tank and regulator from an ad in the local paper.

I brought my equipment into my dorm room, closed the door and assembled it. Having never had a SCUBA class, even assembling it correctly was a hit and miss affair. But eventually, I had everything adjusted. Finally, I strapped it all on (including the fins), then sat down on the floor and, leaning up against my bed, I took the regulator's mouthpiece between my lips and sucked. Nothing happened. It was not only profoundly disappointing, it was kind of scary to suck against a mouthpiece that was supposed to deliver air and get nothing but a hard walled vacuum. (I would have the exact same experience again years later in the ocean when my tank ran out of air – but that's another story.) Soon enough I realized I had forgotten to turn my air on. I twisted the valve, sat back down and drew my first breath of air from a life support system. I clearly remember what an amazing sensation it was!

A few days later, I took my gear out to our local community lake, walked into the water at the end of the boat ramp and sat down on the bottom with the water swirling some six inches over my head. The water was frigid and the act of sitting down caused leaves and mud to whirl around me so that I could barely see my hand before my face. But on that day, I entered the underwater world with my own life support system for the first time. I sat there for as long as I could in the chilly water, breathing the sweet air through my regulator. Even while I was extremely cold and in relatively unpleasant surroundings, it was a truly addictive experience. Fortunately, just a few years later, I would be certified in the relatively warm and beautifully clear waters of Hawaii.

I still love the feeling of taking a breath of air out of a clean, new, finely adjusted regulator. To my mind it is the act and perfect mating of man and machine. And it is a wonderful machine that not only gives air, it also unlocks the door to an undersea world that before was sealed firmly shut to mere air-breathing humans.

An undersea habitat's life support system is one that provides a secondary environment that will allow a human to survive when the primary environment around them may no longer provide those needs. For example, basic SCUBA gear offers the ability to breathe air under water. But the SCUBA gear alone does not also provide protection from cold or even from oxygen poisoning under higher pressures. Therefore, a life support system is made up of many different elements and comes in many and widely varied designs and with diverse expectations.

A basic SCUBA setup may provide air for a half hour dive in cool waters at an average depth and therefore present everything a diver needs for that short slice of time. But if the diver is in the water for a longer time, his body may not be able to keep up with the heat loss. Of course, the tank will eventually run out of air, so that the rating of the system is exceeded and the diver will then die. In this example, the life support system was well designed and operated perfectly – until its limits were exceeded.

If we expand the idea inherent in a SCUBA setup to an undersea habitat, the thought is the same, only the diver is encapsulated in a shell of air underwater instead of being in direct contact with the sea. But the life support system still has to provide oxygen (in the form of a breathable air mixture) and protection from the elements as well as a way to confine and dispose of his waste gas, carbon dioxide. If the aquanaut stays in his habitat long enough, then he will also need to dispose of his bodily wastes, stay warm or cool, control the humidity for comfort and have access to food, sleeping arrangements and, at least for now, eventually have a plan for returning to the surface, which may or may not include a decompression event.

As you can immediately understand, the habitat experience is far more complex than a simple SCUBA outfit, with many more variables included. And yet, it is because of those variables that the habitat opens a very long term window into living and working in the sea that SCUBA can never provide.

In the first generation habitats, there were fundamentally three different kinds of life support systems (LSS):

Constant flow air: In this simplest form of all LSS for undersea habitats, air is supplied in a constant stream from an onshore or barge mounted compressor. In this way, far more oxygen is supplied than the aquanaut needs and, if the flow is high enough, his carbon dioxide is flushed away so that no scrubber is required. The advantages of this set up are many: no relatively dangerous elemental oxygen and mixing system is required; there is no cost of scrubber chemicals; no hazardous waste generated; and no complex LSS monitoring system is required. The disadvantages are that if the compressor fails, the mission is over, and air compressors use a high amount of energy.

Oxygen make-up with scrubber systems: In these LSS, elemental oxygen is mixed with the air stream as it is used by the aquanauts while the CO_2 is absorbed, generally by chemicals. This arrangement uses far less gas but is complicated by requiring systems to be continuously monitored as well as adjusting O_2 and CO_2 levels. It also requires make-up air gas banks to make up for absorption in seawater, leaks and losses. And it has a tendency to hold dangerous trace organic gasses, which if not adequately filtered, can cause problems with human physiology. This system is essentially a whole-system rebreather.

Mixed gas systems: Rather than utilizing surface equivalent air mixes, some early habitats used mixed gas LSS to lessen the toxic effects of oxygen under high pressure and control decompression time when flushing nitrogen out of human tissues. These mixes were sometimes relatively exotic and required extensive expertise for both system design and safe operation. These rather sophisticated LSS were also one of the elements that sealed the fate of the misdirected path in undersea colonization, since it was truly an unrealistic approach to day to day human living and is even difficult and risky for hazardous working dives.

It is almost certainly true that the 'tough guy' tool pusher divers will violently disagree with this assessment, but they would quickly have their minds changed if you would simply ask their wives if they were interested in spending a single weekend in that environment or taking the kids along for family day. Besides, it is particularly difficult to take things seriously when you are being screamed at by a 300 pound, burly, no-nonsense work diver with an unintelligible voice that sounds a lot like Donald Duck.

Now having said that, mixed gas diving has had, and will always have, its clear place in undersea exploration as long as professional work crews and scientists engage in this often risky and dangerous activity.

WORLD RECORD DIVER & AQUANAUT TERRENCE TYSALL

I have been particularly blessed with the friendship of Terrence Tysall (a former US Navy Seal turned Army officer serving in Iraq in 2007) who has established several deep diving world records.   One of those includes a dive to the *Edmund Fitzgerald* where Terrence placed a memorial plaque on her hull to commemorate the lives lost on this famous shipwreck.

Terrance is also the founder of the Cambrian Foundation, a non-profit, scientific research and marine conservation group that travels around the globe conducting research projects, education seminars, preservation activities and exploratory expeditions.

Terrence is one of the most savvy, most experienced and most knowledgeable divers in the word.[9]   I described my preference for protecting the human on the deep dives he routinely engages in by the use of hardsuits, submarines with lock-outs, and/or various other methods.   Terrence smiled and responded with a hard reality – that there are no 'affordable' hardsuits available for teams of professional divers today.   He also pointed out that there are currently no existing working techniques or hardware easily available (or within most private teams' budgets) that could possibly meet the needs of the myriad of undersea tasks accomplished all over the planet on any given day.   He said that if

---

[9]Terrence Tysall is the only human to have dived to the Edmund Fitzgerald using SCUBA.   The wreck is located at 533 feet on the bottom of Lake Superior.   His eight minute visit to the ship cost over four hours of decompression on the way back to the surface.   Terrence is scheduled to be one of the three crewmembers to establish the 80 day world record for longest uninterrupted stay underwater in the 2009 Atlantica I Expeditions.

there were methods and equipment that would offer him and his working colleagues more protection, to, "...bring it on! We would definitely never refuse them!"

In a conversation with Chris Olstad, I warned him that on a particular dive which we were planning, the visibility could drop to near zero. He immediately laughed and said, "Well, Dennis, will it be below the mud line?"

I thought he was joking, but I responded with a smile that no, it would definitely not be below the mud line! (I did not have the heart to tell him that if it was, he would definitely be on his own.)

He then told me that as a professional diver he had been involved in a job where the divers were required to literally sink down below the mud line – into the watery ooze, well over their heads in mud, in total blackness – and cut pilings with torches in zero viz with all work being done by touch alone.

As I listened to him, my skin crawled and I was reminded that the work that professional divers often do is not only totally amazing but it is also work that Aunt Miriam and the kids would never even think about getting involved in under any circumstances!

It was never the purpose of this book to criticize the essential tasks carried out by the diving community who use mixed gas diving techniques. But it is very much my purpose to take the position that these activities will not be a part of the general population's daily procedures or expectations in an undersea community. It is highly possible that mixed gas diving may be necessary early on in the history of the establishment of these communities and with their maintenance teams, but eventually all this activity may well give way to hardsuits and the priority of protecting the human.

The eight basic elements of any undersea life support system include these essentials:

Oxygen Supply: Humans require around 2 pounds of elemental oxygen per day. This gas can be supplied in elemental form and mixed in an inert filler gas or mixed in air (air contains 20% pure oxygen).

Carbon dioxide removal: Humans generate about 2 pounds of carbon dioxide per day. A 'safe level limit' of CO2 in the air has been established by the U.S. government at half of one percent – or 5000 parts per million (ppm). The background level in fresh, outside air is around 400 ppm.

Trace gas removal: Nearly every object has the capacity of off-gassing trace amounts of complex gasses – especially items manufactured by man. These gasses, even in the parts per billion range, can cause problems if they are not filtered or flushed away.

Temperature control: Any life support system must control temperatures so that they are comfortable and safe for the human to live and perform in. The Occupational Safety and Health Administration (OSHA) recommends temperature control in the range of 68-76° F.

Humidity control: Humidity works together with temperature to determine relative comfort zones. OSHA's recommendation of temperature is linked to a humidity control in the range of 20%-60%.

The life support system control mechanism: All these systems have to be controlled in some way. The most basic system is to have an aquanaut monitoring gauges and responding by turning valves. More complex LSS utilize automatic monitoring and system response based on computer controlled algorithms.

Food Supply: All LSSs must include a food supply for the aquanaut.

Resource Recovery: Formerly called 'waste processing', an aquanaut will require the removal and processing of his waste products. Ideally, these will be recycled and reused.

No matter what kind of life support system is assembled – and there are a nearly infinite variety of possibilities – these eight elements must always be addressed.

Alien Atmospheres

Again, one of the central tenants of this book has been to project the philosophy that in every exploration system we must require the systems we build to adapt to the human standard rather than expect the human to adapt to the machine or the environment – and in every design activity, we protect the human as a paramount objective.

But there is a difference between making intelligent choices based on observed results and making rigid rules that have no basis in experience. While it has been found by the early undersea explorers to be totally impractical to live day to day breathing various exotic gasses that may be necessary and adequate for extreme work environments, it would be equally foolish to toss the baby out with the bathwater and force humans to live at sea level standard for the sake of tradition or fear and trembling. And while it has been found that human physiology can pay a permanently negative price for being forced into relatively extreme high pressure environments for prolonged and/or repeated periods, such is not the observed case for incidental exposures or for exposures to lesser pressures for very long periods of time.

Hence, while the one atmosphere standard is an ideal target, there may be other considerations that will allow for an alteration of this standard for the sake of improving others at no risk to the human condition. The key is in understanding what conditions may be changed without adversely affecting the human. And if it is advantageous to make those changes without forcing the human to adapt to conditions that have proven adverse, then it may be prudent to consider doing so.

For example, there is a very large body of empirical evidence that humans may safely live and work at 1.6 atmospheres indefinitely and return to the surface without decompression. This is a powerful finding that allows habitats to be designed to be pressurized to 1.6 ATM. This lessens the engineering requirements for the structure, lowers the excursion depth for aquanauts working outside and thereby reduces the entire operation's risk incrementally. Yet, even an alteration of the atmosphere inside a human community to 1.6 atmospheres still elicits a fundamental change that must be understood.

If I strap on a SCUBA tank and descend to 21 feet (1.6 atmospheres) and breathe air that was pumped into my tank at the surface, I am breathing an alien atmosphere that has immediate effects on my body unlike that at the surface. In this air, the partial pressure of oxygen has increased from 3.1 psi to 4.96 psi. Nitrogen partial pressure increases from 11.6 psi to 18.56 psi. The lungs are a permeable passageway directly to my bloodstream, which means that I have immediately altered my physiology by those amounts. Yet, at 1.6 atmospheres, there are no known ill effects, even over prolonged exposures. Record holding aquanaut Rick Presley was exposed to 1.6 atmospheres continuously for 69 days and suffered no ill effects.

But as the depth and partial pressures increase, the body begins to respond negatively. Pulmonary oxygen toxicity can cause lung damage with *extended exposure* to a PO2 above 8.82 psi (which amounts to an equivalent depth of only 60 feet!) Again, this effect, while not a problem with divers, can become a real threat to aquanauts and is sometimes also referred to as 'chronic' oxygen toxicity. Oxygen convulsions may occur if the PO2 exceeds 29.4 psi.

Likewise, at 1.6 ATM, nitrogen produces no known ill effects. The negative effect associated with nitrogen is called 'nitrogen narcosis' and its symptoms begin with some divers at a depth of around 100 feet and will render a diver unconscious at 300 feet. Remember – these effects are not from some 'exotic gas mixture'. All of these effects are from breathing regular air – whose physiological characteristics change rather radically under pressure into a true alien atmosphere. As you can clearly see, garden variety air becomes an alien atmosphere even at relatively shallow depths.

This is the single reason that the exotic gas mixtures were used from the outset. Because air had seriously deleterious effects at relatively shallow depths, diving physiologists swapped one gas for another in an effort to cancel out or minimize those issues. The problem was, in many cases, they only served to swap out one problem for another.

And, yet, in all of the experimentation, many other significant results were discovered that did not result in any negative effects. Again, prolonged exposure to air in up to 1.6 ATM has, thus far, not exhibited any documented harmful results. Likewise, the slight increase in the PP of O2 at this pressure has resulted in anecdotal accounts of faster healing and a sense of well being. Hence, it would be difficult to apply the one atmosphere standard to this case, since the evidence to date has demonstrated that 1.6 atmospheres may still be inside the margin of what is to be considered 'standard' to the aquanaut.

Therefore, at 1.6 ATM it is thus possible to adjust our gas mixture and pressure to our advantage with few, if any, documented disadvantages. At this level, the human remains protected and the aquanaut is not forced to adapt to any potentially long term harm, even though air at 1.6 atmospheres does admittedly alter the human physiology to some degree.

One of the first gasses to raise its ugly head in an undersea habitat and make its presence known is carbon dioxide. I have spent time in habitats that provide breathing oxygen and flush out unwanted carbon dioxide by simply providing a constant flow of air to the habitat. If all I was concerned about in these situations was oxygen, the air flow could have been reduced from a flood to a trickle. But in order to provide a safe level of carbon dioxide removal, the air flow was turned up much higher to roughly 5-7 cubic feet per minute per person.

Carbon dioxide gas is a waste product of human metabolism. As each of us breathes out, we exhale about 35,000 to 50,000 parts-per-million (ppm) of CO2 – some 100 times higher than outdoor air. If left unattended, this gas can quickly reach undesirable levels. If left alone long enough, a physiological condition called hypercapnia ensues.

On my first experience in an undersea habitat, there were five of us gathered in a large habitat to discuss a mission. After about an hour, right in the middle of a conversation, I suddenly had a powerful impulse to leave the room. My heart was racing and I knew without measuring it that my blood pressure was elevated. But my desire to exit was not logically coherent. I wanted very much to be there. I did not feel any fear, but my skin was suddenly crawling with claustrophobia. All I wanted to do was stand up and move out of that compartment.

I excused myself politely as though I were going to the restroom, slid open the door to the moonpool area and closed the hatch behind me. In a matter of seconds, I felt totally normal. I sucked in several lungfuls of fresh air, composed my nerves and stepped back inside. I sat down and rejoined the conversation, but within ten minutes, the serious discomfort returned. I then excused myself and left the compartment. Again, the same result occurred. Several breaths of fresh air later, things were totally normal again.

But this time, when I returned, one of the experienced aquanauts looked at me and said, "I turned up the air flow." Although no one else seemed to be experiencing its effects, he recognized CO_2's effect on me. Thereafter, I became the official 'CO2 canary' as I was nearly always the first to feel its effects and react to its symptoms.

I have since learned that claustrophobia is the body's defensive mechanism against hypercapnia. It makes us want to toss the covers away from our face at night or escape situations of tight confinement. The body's survival mechanism is to get out of an area in which CO_2 is building. It is common to everyone to feel that urge to escape as the CO_2 builds.

If you decide to 'tough it out' and stick around in an environment of rising CO_2, clinical hypercapnia sets in and things go downhill.   The first effects are elevated blood pressure, dizziness and nausea, but also include gross fatigue and, after a long enough chronic exposure, even paranoia. Eventually, if the levels continue to rise, it leads to loss of motor functions or possibly replacement of oxygen and asphyxiation.

Fortunately for all of us, the body has already worked out a mechanism of clear warning. It is very important for each aquanaut to recognize the symptoms and do something about them before clinical hypercapnia reduces the ability to adequately respond. In professional exploration activities, there is no place for the macho 'I can handle this' tough guy. This attitude endangers everyone's safety.

<u>The three rules of life support are:</u>

Know your equipment – monitor its function.
Know your limit – stay in your limit.
Know your responses – do something about them.

<u>And if I may add this codicil:</u>
In undersea exploration there are bold guys and there are old guys, but there are no old, bold guys.

The Next Generation

In the next generation (NextGen) habitats, the life support systems will be considerably more complicated than the earlier LSS.   As in every other engineering scheme, systems tend to become more complex as engineers integrate new system paradigms to meet discovered solutions to various challenges. The best example I can point to are the automobiles I have owned. When I was in high school my father owned a 1962 Ford Falcon. Compared to my new Suzuki SX4, the Falcon must have had no more than five moving parts! (Please allow me to invoke 'author's license' here.) I joked many times that the Falcon was so basic I could stand down inside her engine compartment and work on her.   The obvious bad news is that my up-to-date Suzuki's integrated systems are so complex, it has passed quite beyond my capacity and knowledge to work on her. On the other hand, on the Suzuki's last day – when she is worn out and I want to trade her in – she will run better, be more fuel efficient and still be more reliable than the Falcon was when my father drove her new off the showroom floor!

Undersea life support systems are just that way. If one compares the systems we have now to the systems we will have in a decade or so, they will become incrementally more complex, and that is a certainty. But they will also be far better, more reliable, more redundant systems than we now know.

Redundancy in LSS is vital. When NASA sent the Apollo astronauts to the moon, all critical systems featured triple redundancy. In other words, when the astronauts completed their exploration of the surface of the moon and pressed the 'go' button on their Lunar Lander instrument panel to rocket safely back into lunar orbit, if the first button did not work, they had a second button with totally independent fire control. And if the second button did not work, then they had a completely independent third system!

Likewise in undersea habitat LSS, redundancy will be essential. If you have an entire colony of hundreds or thousands of people committed to the undersea community, it will not be acceptable to lose power for any number of minutes. In the ocean, as in space, power is life. Hence, all power systems will be backed up and redundant.   Critical systems will almost certainly be designed with triple redundancy.

System reliability is defined as: "The probability that a system, including all hardware, firmware, and software, will satisfactorily perform the task for which it was designed or intended, for a specified time and in a specified environment." Reliability is essential in undersea LSS. Issues of system reliability will become key to what arrangements are selected and which systems are integrated into the colony's engineered framework.

One research agency was working on very complex life support systems for use in space. In these bioregenerative systems, food was grown using very large stands of plants so that the plants could be used for O_2 production and CO_2 absorption as well as producing food and fresh water. During this research an astronaut candidate was given a tour of the laboratories. He listened very carefully, then he stopped and looked the LSS designer dead in the eye and asked pointedly, "And who is going to take care of these plants and harvest them and then turn all this into dinner?"

The researcher blinked and responded bluntly, "You are," meaning him and the rest of the crew. The astronaut shook his head and responded, "No. I am an astronomer. I will fly to do my job, not become a space farmer. None of us have time for this." He then went on to point out that he intended to take the potato from the hydroponics solution, pop it into the microwave, eat it for dinner and be done with it.

The astronaut candidate was not being obstinate at all. Taking care of a labor intensive and complex life support system that requires constant attention was truly an unfortunate waste of his specific talent, just as he said.

He and I and all of us in our culture are accustomed to automated life support systems. They work invisibly in the background so that we can do our jobs in life. When I want to eat, the greatest part of the task is driving to the store and loading up the groceries that I did not have to plant, care for and harvest while I was busy doing other things. Even meal production has been automated for our culture – it has become the expected status-quo. All we have to do, as the astronaut correctly said, is to pick it up and pop it in the microwave.

The air conditioner system in my home is also a part of our LSS. (If you disagree, come and live in the upstairs of my home in central Florida during the middle of the summer without it.) And yet that air conditioner is not operated by me – I only had to program it one time! It is controlled entirely by a 'smart' digital thermostat that not only regulates temperature and humidity, but does so automatically without any human intervention and is smart enough to change its pattern every day of the week! I hardly ever give a single moment's thought to that system because it is operating invisibly in the background of my home. The same is true of a jetliner or of an automobile – the life support systems are invisible and automated.

Automated systems will also care for the human inhabitants in NextGen undersea habitats. These systems will all be smart systems that are automatically controlled so that the humans can go about their duties in life. Soon, the days of manually cranking valves, watching gauges and calculating gas cycles will all be over as undersea systems for colonies are invented then improved.

Earlier we spoke of a bioregenerative life support system. Again, that is a LSS that uses plants for oxygen production and carbon dioxide absorption. These systems have been intensely researched over the past half century with the expectation that they will someday be integrated into space colonies.   Their chief disadvantages are that they require considerable labor, energy and space.   The relatively inefficient process of photosynthesis (6.6%) does not come without a high energy cost.   It requires light energy to split the water and carbon dioxide molecules to produce sugar and elemental oxygen. This comes at an outlay of the watts required to provide light to drive the reaction. Further, these lights produce excess heat which then has to be removed.

Using a dwarf crop stand, it will require about 15 square meters of mature crop to produce enough oxygen for one active person or two people at rest.   That area will provide roughly the same CO2 absorption capacity.   It will also provide about half of one person's daily food requirement – but all at a very high energy and direct labor cost.

Algae systems were heavily researched in the 1960's but this work was largely abandoned after it was determined that algae made a poor food product and astronauts were probably going to justifiably disobey any direct command to eat it anyway. However, algae systems are not limited by canopy size and thus are much more space efficient.   Because of their space efficiency, they are also more capable in their use of light in that they can be exposed to a more concentrated light source.   Further, they are an aquatic species so that their medium can act as one side of a heat exchanger.   All in all, algae seem to be an ideal component for further investigations in undersea colonies.

As was previously mentioned, the 2007 mission of Lloyd Godson in his Australian *BioSUB* experiment was the first known use of incorporating an algae system into an undersea habitat's life support system.   His algae system provided about 10% of his oxygen needs and the same amount of CO2 absorption.

Undersea colonies will evolve from simple LSSs to advanced LSSs and eventually hybrid LSSs. A simple life support system is one that provides oxygen, absorbs or disposes of carbon dioxide and packs and stores food onboard for the mission. It also collects and properly disposes of wastes and trash.

An advanced LSS is one that recycles biological components and produces its own food. A truly advanced LSS is totally self sufficient and recycles everything. They can be either based on the bioregenerative model or the physiochemical model where every recycling step is accomplished with physical and chemical means only and no biological intervention. This technology is so advanced that not a single functioning complete advanced life support system has ever been built for ocean or space applications.

An Advanced Controlled Ecological Life Support System is one that uses purely bioregenerative systems. A CELSS system is completely controlled by bioengineering processes and is balanced by human intervention.

An Advanced Biospheric Life Support System is one that is very large. It is necessary to be so massive in order that the biological components (or biomes) can come into balance with one another and begin a natural cycle with little or no human intervention.

A hybrid LSS is one that has a few or many components of one or another system. Lloyd Godson's simple but elegant underwater life support system is an example of a hybrid system. It used many methods to achieve his LSS goals – from injection of air as his O_2 carrier, to absorption of CO_2 into chemical mediums as well as using his algae system for partial O_2 production and CO_2 absorption.

Almost all systems of the future will be hybrid systems. They will be set up as 'breadboard' type systems that have the capacity and versatility to plug into and integrate LSS components individually into the cohesive working whole. For example, one such system could incorporate an algae system that utilizes space lighting. But some of the life support system's oxygen may come from surface supplied air, while other air is supplied by a pump on the surface that is powered by passing ocean swells. The habitat's carbon dioxide can be absorbed by a seawater exchange pump (described later in this chapter), as well as a chemical scrubber and the algae system, of course.

All habitat life support systems will require gas exchange. As was mentioned before, every human onboard will require about two pounds of oxygen per day and will, in turn, breathe out approximately two pounds of carbon dioxide. Supply of oxygen can occur by several established methods. It can be supplied by bottles of compressed elemental oxygen, by compressed air or by a plant system. As in every human enterprise, cost will probably drive the solution. Since we are discussing permanent human colonies, a large volume of oxygen will have to be provided each day.

It is assumed that early colonies will have a surface connection of some form, described in this book as the S3 – 'Surface Support System' – whether it is a large artificial island where supply ships can dock, a platform, barge or a simple buoy floating on the surface above the colony. At this surface connection, there will be a connection that will be able to send air down to the habitats of the colony. In small, first generation permanent colonies, this air flow may be supplied by compressors that are switched on once per day to recharge habitat cylinders. Oxygen can be supplied in the form of compressed air, cryogenic supercritical air or liquid oxygen. Depending on the size of the colony, air component cost will drive the selection.

Carbon dioxide removal is a separate engineering issue. We breathe out carbon dioxide with every breath as a result of normal human metabolism. As I sit here writing these words, I am exhaling carbon dioxide at a rate of about 0.3 liters per minute. If I am hard at work, I may exhale as much as 1.5 liters per minute of CO_2. But the amount of CO_2 I produce over one minute here at my desk does not really concern me. If I open the window and allow the outside breeze to cleanse the air, it would not matter how long I sat here. But if I sealed my study, the CO_2 build up over an hour or a day definitely would matter!

On the surface, we control CO_2 in our environments by circulating fresh air. But undersea, there is no fresh air to circulate! However, in the early 1990's, Aquanaut Chris Olstad conducted experiments based on a remarkable discovery made in 1963-1965 by H.P. Vind, an engineer at the U.S. Naval Civil Engineering Laboratory. Working from Vind's findings, Olstad set up trays in the *MarineLab* habitat that drew water from the ocean and circulated it though the habitat and back into the ocean. He was able to expose a surface area of 55 square feet to the atmosphere of the habitat. He then shut the air flow to the habitat down so that the only CO_2 removal was from the circulating seawater that came into contact with the air being dispersed and mixed by fans. By this method, he discovered that the seawater alone provided enough CO2 removal for two aquanauts at rest for short exposures, and a single aquanaut at rest for periods of extended duration![a]

This process has been further investigated by Naval Academy Professor M.L. Nuckols in NOAA's *Aquarius* habitat.[b] This single discovery alone may be one of the most important breakthroughs in the history of aquanautics, because it helps eliminate one of the most expensive and formidable barriers to permanent habitation of the seas. While it has been known for decades that ocean water plays an important role in the removal of CO2 from the atmosphere in a process called sequestration, this is the first time it has been applied to human habitation in undersea colonies. Improvements in these seawater CO2 scrubbers have only just begun.

Further experiments are being planned by me and my crew in the *Leviathan* habitat. A smaller, more portable version is also planned for the DST II Research Submarine for its extended seafloor operations beginning in 2010. It cannot be overstated that this discovery represents a remarkable and very important advance in permanent habitat systems. It is likely that these systems will be integrated into all future habitat life support systems because they use an *in situ* resource – seawater, for which there is an endless supply at only the cost of the energy to pump it – as a scrubbing medium!

How effective these methods will ultimately become remains to be seen as investigations continue. But as an absorptive medium, it is possible that seawater will also be able to act as a localized medium supplied scrubber, possibly removing trace organics as well!

CO2 can also be removed through pumping large amounts of air through the habitat. This method is energy intensive and also requires that the habitat be pressurized to ambient depth. Due to these limiting factors, it is unlikely that this method of CO2 removal has much of a future for deep water habitats. However, in one possible engineering process, compressed air can be pumped to a one atmosphere habitat in the oceans, the pressure reduced before entering, then dumped though a rigid hose back to the

The Leviathan Habitat
Designed to investigate systems and techniques for permanent undersea colonization.

surface. While this process has the same effect, it, too, uses a great deal of energy to provide O2 and remove CO2. As in all other processes, the relative economic and hazard comparisons will ultimately drive the choices.[c]

When considering the various economies, CO2 can be absorbed on traditional chemical substrates such as baralyme, lithium hydroxide or advanced resins. However, this can prove to be expensive as well as generate a large amount of waste. At today's market cost, if scrubber material is used as the sole CO2 absorption medium, at an average CO2 production rate, given a mid-cost scrubber (not Lithium Hydroxide) and average activity level, it would cost around $35 per day to purchase the material per person. If the mission utilized four aquanauts for 30 days, the total cost for scrubber chemicals would be $4,200. Likewise, an 80 day mission would cost $11,200.

As a comparison, the cost of the top grade scrubber Lithium Hydroxide would cost $50,000 for the same mission! In a permanent station with 10 aquanauts, Lithium Hydroxide scrubber materials would overwhelmingly cost over half a million dollars for a single year of operation and they would generate a barge full of spent materials. Obviously, because of the significant size of these numbers, it would be most advantageous to engineer alternate methods of CO2 absorption whenever possible!

As we consider each of the LSS requirements, perhaps one of the most important design concepts is in colony placement, because where the colony is placed determines many different aspects of colony and LSS design. Placement ultimately even defines the colony's energy budget. One such aspect of habitat placement is the prevailing temperature of the waters in which it is located.

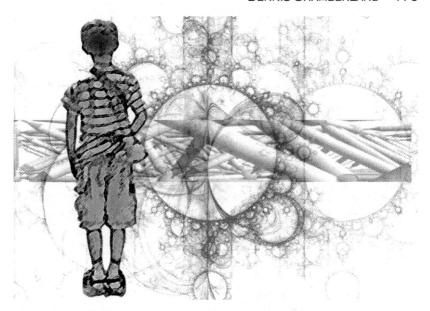

COLD AND WARM

"... those who have once listened to the siren songs of the
ocean bed never return to land."
Philippe Diole.

O f all aspects of direct human contact with the undersea environment, perhaps none is as important as temperature. The human being is comfortable in air temperatures of about 72° Fahrenheit with humidity at or below around 65%. But in water, unless the temperature is around 82° Fahrenheit or higher, the water will begin to sap the heat from the body at a rate that is not comfortable over relatively longer periods of exposure. All bodies are not created equal, of course, and some humans are far more sensitive to this effect. Individuals with more of a fat layer (such as warm blooded aquatic mammals like the seal, and more corpulent humans) will tend to hold in their body heat considerably longer than humans without much of a fat layer because the fat acts as a heat insulator.

The human body is thermo-regulated around 98.6° F and at least 70% of the body's content is liquid. Therefore, water below 98.6° F immediately begins to remove heat from the body at a rate

25 times more efficiently than from air. Interestingly enough, at least half that heat is lost through the head! The body will shut down the capillaries to the extremities first when heat is being drained from the human, but the head is the last to shut down – hence the higher than average heat loss through that part of the body.

Further, added activity in the water increases heat loss, so that survival time is decreased if the human is briskly swimming, swimming in panic or for survival. Physically powerful swimmers have actually died swimming 100 yards in cold 50° F water. And in water under 40° F, victims have been known to perish before swimming even 100 feet!

As the body's temperature begins to decline, autonomic, protective functions kick in. At a body temperature of between 97 and 90° F, shivering commences. About this point, gasping begins, which in itself may cause a loss of consciousness due to hyperventilation. At 93° F, mental dexterity is lost and helplessness sets in. When the body's core temperature reaches the mid 80's, unconsciousness occurs. Death thereupon follows either from drowning or when the core temperature drops below 80° Fahrenheit.

This information is presented because the *average* world ocean surface and near-surface waters temperature is about 62.6° Fahrenheit, far lower than the safe and comfortable range for direct skin exposure. At latitudes near the Polar Regions, the surface waters and near surface waters can be as frigid as 28.4° F although the equatorial latitude waters may be as warm as 96.8° F. Further, most of the ocean below the sunlight zone is between 32° and 37.5° F! As is shown in the accompanying graph, water temperature declines asymptotically (has a straight line that a curve approaches progressively more closely but never reaches) with respect to the freezing limit.

This book is seriously considering human habitation of all the undersea regions of the earth, and since most of those areas are below the comfortable and safe zone for direct human skin exposure, then it is reasonable to expect that I will also recommend countermeasures to this rather adverse environmental condition.

The countermeasure is obvious and has been presented

repeatedly in this work: protect the human!   Humans are protected from cold water first by wet suits of various thicknesses, then by dry suits which insulate the skin with a layer of air and finally by hardsuits which protect not only against temperature extremes, but also against pressure.

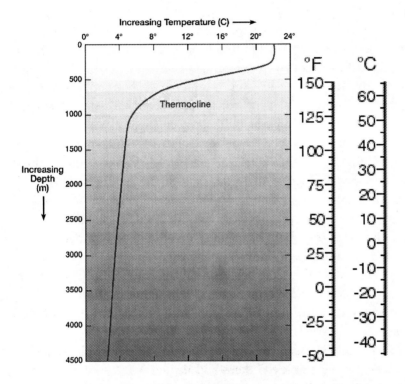

When viewing the photos of the astronauts walking on the moon, I have never even once heard someone remark, "Gee, isn't it terrible that he couldn't walk out there in his Speedos?"   Why would anyone never actually make the remark?   Because the statement would be recognized as supremely ignorant and foolish. Even the youngest child understands that spacesuits protect the human against adverse environments.   And yet, due to the widespread   misunderstandings   engendered   by   the   early misdirected   starts   in   undersea   exploration,   there   is   some expectation that humans can weather just about any form of underwater discomfort by willfully offering bare skin directly to the sea.

I am told by my companions that have been diving under icebergs and glaciers that every patch of uncovered skin burns and tingles when exposed to this frigid water. Exposing larger areas immediately endangers the aquanaut, not to mention the instantaneous misery it engenders.

In one of maritime history's greatest sea tragedies, the sinking of the *Titanic* after striking an iceberg on a sideways glance, most of the 1,523 passengers who perished did not die from drowning, but instead from the direct exposure to the agonizingly frigid North Atlantic waters.

In the next generation of undersea habitation, much of the ocean environment will be considered just as alien as the space environment and humans will protect themselves against it – especially in waters with lower temperatures which make up most of the undersea regions of the earth. Obviously, the ambient temperature will drive many aspects of the habitat's design and will determine the colony's overall energy budget. Habitats in tropical and subtropical zones have no need to calculate heating into their designs. But, just as land dwellers know, if heating is to be an issue, it can be a significant energy drain.

Undersea habitat heaters will, of course, be from electric resistance and similar type heaters. No flame heating is practical due to its consumption of oxygen and production of unwanted gasses as a byproduct of the process. However, if there is a large and versatile S3 (Surface Support System), it is possible that flame heaters may be employed there and heated water sent below in an insulated umbilicus.

In *SEALAB* II, which was placed in the cold waters off Southern California, heating cables were positioned low to keep the habitat's deck warm with hot water services provided by the large, 110 foot surface ship tending them above. Yet, in permanent operations, it is very unlikely that most early colonies will be able to afford such a luxury as an S3 platform that is manned full time by a crew of dozens, operating such systems for the residents below. In fact, as we will discuss in a future chapter, most S3's may not necessarily be manned at all! Under any circumstance, heating an undersea habitat will be expensive in terms of energy use. Therefore, it is essential that habitats designed for cold water be adequately insulated.

Habitat insulation is much more important than in a human surface home. Habitats with metal walls are particularly vulnerable to heat loss. Indeed, their walls become a heat exchanger directly with the water outside. Like a whole system radiator, the habitat will radiate its expensive heat right out into the sea unless it is properly insulated.

Properly positioned insulation may include a layered boundary, so that the insulation material is not placed in direct contact with the skin of the habitat. In this design, the habitat's insulation is placed in a layer one to several inches from direct contact with the walls. Air from the habitat is allowed to circulate between this layer on its designed path through the life support system in order to utilize the cold skin of the habitat to dehumidify the air instead of relying on a coolant loop, which is another expensive energy cost. Hence, in doing this, the habitat can effectively utilize its own walls to dehumidify the passing air stream at no cost! Of course, the annual ambient temperatures of the waters would have to be known before deployment. If the ambient water temperature rose higher than the habitat's air/water condensation temperature (dew point) at any time during the year, an alternate dehumidification system would have to be added.

Habitat insulation is also unlike that found in homes in that it must be biologically inert, not allowing molds and other microorganisms to begin growing on its surfaces since it is in these areas that these biological layering and biofilm problems would most likely originate. Likewise, habitat walls that are used for condensation of water vapor must also be pre-treated with anti-fouling paints to prevent excess accumulation of biofilms.

Humidity control is one of the most critical issues involved with any habitat design. And, of course, humidity control is invariably associated with ambient and internal temperature profiles. In warmer waters, humidity can be controlled with onboard air conditioning systems. In colder waters, as was previously discussed, perhaps the walls can be used for condensation and dehumidification. In the case where this is not possible, a separate dehumidifying system must be installed.

As humidity levels climb, any living environment becomes less and less pleasant. In the case of a breakdown of humidity control, under certain conditions, it can become so incredibly miserable that it may actually become necessary to leave the habitat, as I have personally discovered. When I was operating a habitat in a sub-tropical lagoon in the summer, my air conditioner system failed. I lost both cooling and dehumidification at the same moment. A technician immediately responded to repair the problem, but in the 90 minutes it took to solve the issue, I was beginning to consider evacuating the habitat. The conditions were so miserable that I believe they were nearly intolerable. Yet, once the small air conditioning unit was switched back on, within minutes the habitat became quite livable again.

In a cold water habitat, if the humidity control system breaks down, the water vapor in the air will begin to condense on every surface, causing the habitat to quickly become clammy and damp. Whether the water outside is warm or cold, either way, there is no getting around careful planning for humidity control in the design of every habitat.

I am now designing a four man habitat for an extended underwater mission in sub-tropical seas. It is my intent to install three completely redundant air conditioning and dehumidification systems onboard. Because of my past experiences, I recognize that dehumidification is just as mission critical as any other life support function.

Another issue with cold water is window condensation. Windows can be insulated by double layers, of course. But windows that are in direct contact with cold water will immediately fog and begin the process of condensing the humid cabin air. These kinds of problems are very serious for several reasons. One: visibility outside will be more or less permanently obscured. Two: a permanent condensation stream will create myriad problems identical to a constant leak in the area. Three: the condensate will end up a unending pain in the rear that has few good solutions.

It is highly recommended that all windows in habitats to be placed in waters below 60° Fahrenheit be double layered with a circulated dry air boundary between the layers. If these windows are not properly designed, the inner layer will condense and there will be no way to wipe it clear! If all else fails, a circulating fan or air blowers can be mounted at each window. The lower the humidity in the cabin, the less a problem any of this will be. However, habitat humidities less than 20% will generate yet another list of problems.

In habitats without insulation in cold water, condensate will form and the skin of the habitat and all the walls, as well as other exposed surfaces (including bedding), will become wet and stay wet. This is perhaps one of the most miserable of all situations. Eventually, the wet walls will absorb trace amounts of carbon dioxide and thus create a ready carbon source for biofilm that naturally occurs on all surfaces on land or undersea. Within weeks, the walls will become slimy as these biofilms grow. In short order, these missions will end and the habitat will have to be evacuated. I have heard a highly placed rumor that such an event actually occurred in a Soviet Salyut space station before it was de-orbited.

Habitat walls act as radiators one way or another, and habitats placed in warm water still deal with the same physics. In the case of very warm waters, the habitat will quickly radiate the warm, outside water temperature to the inside as the habitat system seeks thermal equilibrium. If the water temperature is very warm, this can become uncomfortable and have to be dealt with. As I have repeatedly stated, warm water and humidity team up to cause habitat conditions to become unbearable in short order. One may stubbornly overcome the misery for hours or even a day or so, but eventually, the mission will end early.

The best way to meet this challenge is, again, good insulation so that whatever thermal control systems are at work inside the habitat can maintain comfortable conditions without temperature being easily lost to the outside. In warm waters, a standard air conditioner system may be used to not only cool the air, but, in the cooling step, to dehumidify the air at the same time. It is a perfect and elegant double role solution to both cooling and dehumidification.

In the *MarineLab*, Chris Olstad and his aquanaut team purchased a small off-the-shelf air conditioner, cut the outside heat exchanger coil off and replaced it with a copper coil. This loop ran outside the habitat and back into the air conditioner inside. They then recharged the unit and turned it on. It has worked beautifully for more than a decade, providing all the cooling and dehumidification required.

THE SCSAS EXTERNAL
HEAT EXCHANGER

In the *Scott Carpenter Space Analog Station*, we brazenly copied their brilliant and time tested idea and, once again, the system worked flawlessly for two summers of operation in sub-tropical waters! We had all the air conditioning and dehumidification we needed for a total cost (including installation) of less than $200. That simple system worked so well because the surrounding water, even at 90° F, could provide adequate heat exchange since water transfers heat with a much higher efficiency than air.

You would never build a home in Alaska that had the same design characteristics as a home in Florida. The heater in the Florida home is added almost as an afterthought but its air conditioning system is sized for non-stop operations around nine months a year or more. Any fireplace is mostly for decorative purposes. The Alaskan home is designed for far different conditions. It may not have an air conditioner at all. But its heating system is large, robust and designed for eight to ten month annual operations or more and its fireplace is anything but decorative.

Undersea habitats are no different. An undersea habitat placed off the Florida Keys or in the Bahamas will be designed far differently than one placed off Southern California or the Aleutian Islands. Water temperature design is critical in deciding the success or failure of the permanent undersea project, as well as the consideration of lighting.

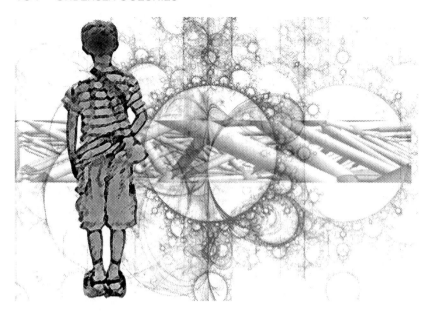

DARKNESS AND LIGHT

"She walks in beauty, like the night of cloudless climes and starry skies; and all that's best of dark and bright meet in her aspect and her eyes: thus mellow'd to that tender light which heaven to gaudy day denies."
Lord Byron - 1788

As I discussed previously, the undersea environment is quite different than that above water, especially when considering light. Even under the very best of conditions, undersea light will always be far less brilliant than on the surface. The human eye is a wonderfully adaptive instrument, and will compensate for dim light by opening the pupil to allow more light to enter, so that under most shallow water conditions the dimmer light is almost unnoticed. Further, the undersea spectrum is very different, so that the change in light intensity is somewhat masked by the spectral changes. But, as you go deeper, as the visibility decreases or the day grows short, the light intensity changes are fully appreciated.

When working outside the habitat, good lighting is a must. Most professional dive helmets have attached lights. I have always appreciated the head mounted lights versus the hand held lights because they free both hands to work. When considering habitat designs, it is important to also consider outside lighting. On one of my habitats, I installed an external moonpool light so that any aquanaut approaching the habitat at night could clearly see their way and would not accidentally bump their head on the moonpool trunk on the way in. I also installed a pair of lights on the front of the habitat so that we could view operations on the forward parts of the structure and so that anyone working at the forward mounted valve tree at night could see what they were doing. These forward lights were also designed to be skewed to the right or left so that the habitat could likewise be illuminated on the port or starboard sides. The external lighting was also the source of endless hours of enjoyment at night as we watched the sea life attracted to it play in and around the beams.

I do not think I can overemphasize the importance of the internal lighting scheme in the design and functioning of an undersea habitat. If the eye is the window into the soul, then light is the key to the heart. Light plays a subliminal role in determining a sense of well being and peace. Inadequate lighting can lead to depression and even anxiety. Knowing that up front is an essential key in design and operations.

The first thing a good habitat design employs is windows. Windows in an undersea habitat are the source not only of free lighting, but also open up the undersea world to the human. The problem with windows is that most habitat designers regard them as areas of unacceptable structural vulnerability. This is a very well placed sentiment. Windows do offer a pair of significant structural vulnerabilities – leaks around the seams and potential structural failure of any window which would lead to immediate, catastrophic flooding and loss of the mission. But in my view, these risks, when properly mediated, are by no means unacceptable.

Part of the negative sentiment, I believe, is subliminally introduced by naval submarines. They have no windows at all. Yet windows have been placed in every structure that has gone to the deepest parts of the ocean, such as the *Trieste* in 1960. Therefore, it is certainly not depth that plays a role in deciding whether windows are placed in military submarines – it is hydraulic shock.

If a submarine is attacked by depth charges, mines or torpedoes, the designers wanted to absolutely minimize areas of vulnerabilities, and the windows went first. Besides, military submarines are not out to enjoy the view as they primarily pass from one point to another in empty and usually dark water. And yet, the image of the windowless submarine is not lost on our psyche. We automatically tend to think therefore that *'submarines are not supposed to have windows'*. Yet, many of those same risks are implied to exist in other undersea structures as well. But I will argue that they do not. If an undersea colony is attacked by submarines, the battle will be lost before the first torpedo is released, so any habitat designer can absolutely scratch 'torpedoes and depth charges' off their worry list!

When I design habitats, I approach windows with a balanced tension: the benefit of the window verses their inherent risk. The risk is real but the benefit is also just as real. A first-rate habitat plan will consider all of these aspects and design accordingly.

I begin a habitat design the same way I plan a book project – with a blank sheet of paper. With that sheet of paper, I can proceed in any direction I wish. I typically make my design 'outline' by defining the general floor plan of the habitat based on mission requirements (number of crew – both permanent and rotating; mission duties; etc.). Once the floor plan is finished, I then construct the three dimensional shell around it. When that is complete, I stand back (typically looking at a rotating computer generated model) and decide the most advantageous positioning of windows and their placement.

I mentally position myself inside the model and look around, imagining where the windows should be; what the view will be; considering why the windows were placed there and what their purposes are. For instance, why place a large window (greater risk) in a place where a smaller window would do just as well? Windows that are strategically placed for inspection purposes have no need to be large. But windows placed for lighting and view should be larger. At that point, I have the freedom to begin inserting windows into my design.

There is much argument against placing any upward facing windows in a habitat design at all. The idea is that any object (such as an anchor) can be inadvertently dropped from above and would destroy a habitat if it struck a window. I argue that an anchor could destroy a habitat whether it struck a window or not. It is true that the window 'may' make such an event more catastrophic, but not necessarily so.

I reason that the upward facing window is also doing something for you day by day – something extremely positive. Hence, rather than making a blanket rule against the design, I would much prefer to try and make tradeoffs for keeping it. The tradeoffs are in upward facing window protection. In one habitat, I had an upward facing dome that was especially vulnerable. But it was psychologically important, offered a 360 degree view and it was a perfect, highly efficient skylight in the main space, so I decided the multiple benefits were worth the risk. I engineered a 3-D cross brace and put a light mesh screen around it to prevent any direct contact. It was a compromise, because the screen had to be cleaned each day and the view was never perfect, but it ultimately preserved the view. In the end, I kept my awesome dome, its light and my peace of mind at the same time.

In the same habitat I had a rather large upward facing flat plate window in the wet room. I protected it with a flat cross brace. It, too, offered bright natural lighting by day. I also used bullet proof polycarbonate for my window material. This material gave me a great peace of mind knowing that in some impact scenarios, the material strength was even more impenetrable than the steel surrounding it.

That particular habitat was an ambient pressure habitat so, if in some bizarre event a window was cracked and the air started to leak out (remember in an ambient pressure habitat water cannot leak inside except from the moonpool up), we had a roll of duct tape at the ready. Not only would the ever-handy tape have sealed the crack in most cases, but it probably would have allowed the mission to continue to completion!

Upper facing windows in undersea habitats are quite rare. In fact, when my habitat was sitting on the sea floor, every experienced aquanaut that came inside just stood and silently stared upward. It was a view that most of them had never seen before from inside a habitat.

There are two kinds of upper facing windows. One is a window slanted at some degree less than parallel with the surface. And the other is a skylight type window pointing directly upward. Both offer surface views depending on where the aquanaut is seated or standing. I particularly like both. From the front console of one habitat where our control chairs were placed, you could clearly see the surface and forward of the habitat. At night, it was awesome to see the moonlight shimmering off waves on the surface or in some cases, see indistinct flashes of lightening. But the chief advantage was that the interior of the habitat was fully and adequately lit on most days by natural light so that artificial light was hardly ever required. This saved both energy to illuminate the space and energy to remove the light's heat.

Another altogether different kind of light is available by using bottom facing windows. Although none of my habitats have ever incorporated these, I have been in undersea habitats that did. Depending on the habitat location, this lighting is a very diffuse and muted illumination that is reflected from the deeper water or from the bottom and into the habitat. It also allows the aquanaut to view bottom dwelling sea life and is a very interesting addition if the design allows it.

I have spent much time in habitats that offer windows only on one or both ends. This does permit the entrance of ambient light, but it also completely obscures the view of the surface. End facing windows also do not permit the brightest light from the surface to enter the structure directly and therefore, end facing windows alone usually require supplemental artificial light to allow for adequate interior habitat illumination. The worst part about end facing windows is the sore neck muscles that come from straining your head to look upward to gain even a glimpse of what lies above. By a careful habitat design, you can wisely incorporate windows that allow ambient lighting from many different perspectives and orientations and, at the same time, permit a more intimate and psychologically pleasing experience with the sea environment.

There is one certainty that always comes with the undersea environment – the illuminated part of each day is always shorter than on the surface. The sun rises and sets with the same frequency as it does above, but the light that makes it into the habitat is always less than the light on the surface, so that in the undersea colony, the day brightens slower and dims faster because of the filtering effect of the water layer above it. And, as I discussed previously, the spectral colors are far different than on the surface.

Habitats placed below 150 feet are always in a kind of perpetual twilight. Habitats placed below 250-300 feet are in near darkness all the time, and the sunlit portions of their days, even with bright sunny skies above, are still very limited. Habitats placed below 500-600 feet are continuously in near or total darkness.

Eventually the time comes to turn on the lights inside the habitat. Or, depending on the habitat depth, the lights may always remain on inside. In most every undersea habitat there are constant lighting requirements in one or more parts of the structure. In all these cases, whether the colony is at 30 feet or 300, interior lighting is very important.

Proper lighting is directly associated with psychological wellness with many people. Key studies have shown this to be true. In the case of Seasonal Affective Disorders (SAD) it has been shown that light intensity is proportional to mood and mood swings. The prescription for treating this disorder (which is typically found in northern latitudes during winter months) is banks of bright lights. Typically, these light banks are required to have an intensity of 1000 lux (a unit of illumination) to be effective. This syndrome cannot be assumed to be limited to a few people. One study suggested that proper lighting is an essential ingredient in every healthy person's makeup.[a] There are federal standards for what the government considers to be proper lighting in the workplace, defined as at least 20-50 foot candles at the desktop level.

In one of my habitats, at the suggestion of the electrician, we installed high efficiency fluorescent bulbs in each space. In the wet room, bath area and main passageway, we installed cool white light spectrum bulbs. That light was considered most appropriate for working environments. But in the main cabin area, we installed yellow spectrum fluorescents with a spectral quality akin to an incandescent bulb – a softer one with a yellowish tint – such as is found in most family living rooms and bed rooms.

When he suggested this lighting scheme to me, I was at first quite skeptical. I was doubtful that most eyes could even detect the spectral variation and I was very convinced that the altered light qualities would never make any actual difference. But I was very wrong! The brighter, white light in the wet room was extremely effective for all the activities that we conducted in there. And in the living areas, the transition from the white light to the more restful, incandescent type spectrum was very relaxing and literally changed the mood from one space to the next. I can truthfully say that this lighting strategy in even such a small habitat made a dramatic difference to the quality of living and working there.

Another decision we made early on was to paint the bulkheads, or walls, a bright white. That choice was made in order to most effectively reflect every available light source as well as to lend the impression of a larger space. In the end, we found it was a very good decision and worked very well, validating both concepts.

I have been in other habitats where the walls were decorated with various insulative materials. In these cases, I have found mixed results. With darker colored materials, it made the spaces appear tight and confining and positively quenched the light. However, where the space lights were backlit and projected onto the walls, the darker colors were softened and blended well with the lights, voiding out the tight and confining feelings.

In all my future lighting schemes, I am going to plan for programmable lights such that every space can command the light intensity and spectrum they require from the command and control computer console. This gives almost unlimited versatility in space lighting decisions which can be made on the spot for any given occasion, or can be changed by anyone in that space whenever they feel like a new lighting scheme. (Do you get this instant impression of impending 'light wars' ahead?) No matter what, I would strongly recommend that before any final color and lightning scheme is installed, if at all possible, a mock-up space be constructed and the elements combined to give some indication of the validity of the lighting plan.

A small, dark space will invariably lead to dissatisfaction in anyone, regardless of their psychological make-up. The days of ex-military men who bravely slip into black steel tubes and then grit their teeth until they adapt to the misery – or until the mission is concluded – are no more. We have now entered a new phase of habitat design that will actually lead to families permanently settling undersea because they live in created spaces that are intelligently and attractively designed – as compared to men who find themselves permanently gritting their teeth at the end of each mission and ultimately end up running into the nearest forest screaming (why do you think Rambo went to live with the monks?).

All forms of lighting are a source of energy consumption. The light fixture must be powered to emit a prescribed amount of energy at the desktop or floor. These, as well as any associated transformers that allow the light to work, create heat. This heat may become waste heat that the heat exchanger is then required to remove at the expense of more energy. Hence, lighting schemes that efficiently utilize energy and produce high levels at low power consumption are very desirable. The energy output of our habitat lights were rated at 75 watts but they actually only used 15 watts of power.

Even more efficient than these bulbs are the recently developed LED lighting schemes. In the undersea habitats that I am currently designing, we will utilize LED light panels throughout. These light panels are the most efficient and lowest waste heat generators available. They are also the most versatile in precise spectrum production and, with proper design, can be tuned to change from one spectrum to another and across a range of intensities by a master controller. In fact, a computer can control these changes either by an automated master sequence or by override commands.   They also outlast the installers and will literally burn 24/7 for decades.

External lighting in undersea colonies will be especially important to give the Aquaticans a three dimensional point of reference and to underscore the magnificence of the undersea environment into which they have purposefully placed their colony. A lighting scheme around each habitat, as well as outlined light patterns that connect each structure in the community, is specifically designed not only to navigate between structures, but to visually unify the purpose and multidimensional framework of the world they have fashioned.

The human being has been created to feast on light, to be attracted to light and even to long for light, as we have discovered, to provide a significant component to their sense of well being. It would be manifestly foolish for us to ignore that and, once again, demand that the human adapt to the darkness. If we maintain that mindset, the population of the oceans will certainly remain zero for a long time to come.

However, with some creative ideas and new architectural designs, Aquaticans will soon have stunning structures to inhabit.

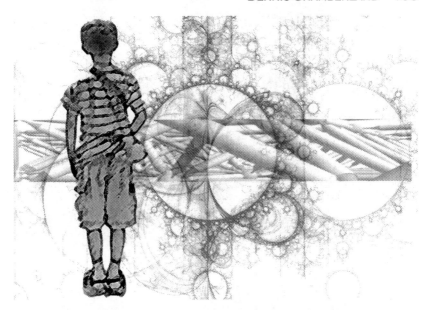

A NEW ARCHITECTURE

"Here's to the crazy ones. The misfits. The rebels. The trouble-makers. The round heads in the square holes. The ones who see things differently. They're not fond of rules, and they have no respect for the status-quo. You can quote them, disagree with them, glorify, or vilify them. But the only thing you can't do is ignore them. Because they change things. They push the human race forward. And while some may see them as the crazy ones, we see genius. Because the people who are crazy enough to think they can change the world, are the ones who do."
Apple Computers Advertisement

While in college, I lived next door to several architects and befriended them. The most well-mannered thing I can say is that I found them to be a rather eclectic bunch. But I carefully listened to their philosophy of artfully blended form and function and grew to respect and admire it greatly. I learned from them things I have never forgotten. And as I have visited city after city around the world, I have seen their craft unfold and impress itself on skyline after skyline.

If you ask a billionaire captain of business how he wishes his skyscraper to be built, it is a certainty that he will not respond that he does not care. I would be willing to wager that not one CEO would reply that they just want a stack of cubes so they can fill it with workers to make them even more money. But I am fairly certain that each of them would want to leave a legacy of beautiful form as well as function to bear their name.

It is buried somewhere deep in the human psyche to want to fuse the beauty of form with the sheer power of function. The movie hero can dress up in a ragged goat skin and go out to defeat the enemy with a club. But more often, in successful movies, the hero will appear onscreen in gleaming armor and cut his enemies to pieces with his beautifully appointed sword.

Does it really matter if the undersea living structure looks spectacular or does it simply matter that is a safe haven? The question is, "Do we want it pretty or do we want it to work?" The answer is, emphatically, "Both!"

If I really have to stand toe to toe with a reasonably intelligent person and justify beauty of form fused to the power of function, then I am wasting my breath. For those who cannot and do not understand this concept, no amount of tutoring will ever make a difference anyway. Sadly, I have had this unenviable task before and have found that no amount of reasoning will ever convince such mindsets otherwise.

If we look back at undersea habitat architecture, it is certain that most of these structures were designed with little thought for form. This was mainly due to cost limitations. Some of it was, unfortunately, a result of little or no imaginative thought. There was one private group in the early 1960's who moved underwater in the interior cavity of a cement truck mixer. Their excuse was fully acceptable: no funding. Unfortunately, the US Navy did not have that excuse when they purposefully built *SEALAB* II and III to look like a railroad tanker car. The only comment I can offer in this case is that I used to work with some of those folks and I totally recognize their mindset. Hence railcar architecture does not surprise me in the least!

Please note that totally understanding them does not mean I have any admiration or approval for the process that leads there!

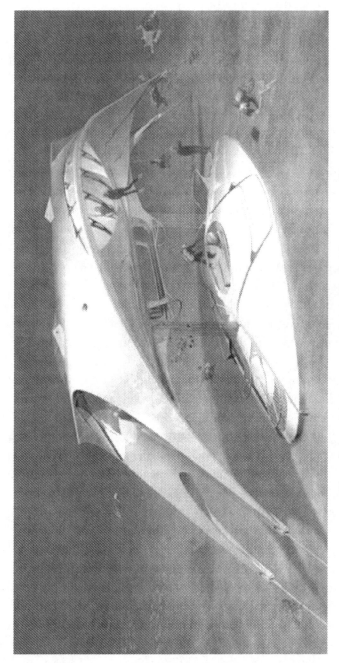

French Architect Jaqcues Rougerie's
Magnificent Maison Sous-Marine — Observatorie
© 2006 by Jaqcues Rougerie

The *SEALAB* habitat functioned beautifully. Unfortunately, it was not a home to which anyone would want to actually bring their aquanaut girlfriend to impress them.

At the same time, half a world away, the Cousteau team was sincerely plugged into the power of fused form and function. All of Cousteau's undersea architecture was clearly well thought out and there was an obvious conscious effort to ensure that his architecture was both powerful and effectively merged with the required function.

Other programs around the world plodded unimaginatively onward in habitats that looked more like propeller-less submarines or barge mounted LP gas tanks. Some habitats resembled submerged oil-rig platforms, and I personally submerged a Lunar Lander look-alike (…just to spread the derision around equally).

However, in the early 21st century, a designer of one atmosphere resort habitats, Bruce Jones, released his view of undersea habitats that were strikingly beautiful in both form and function. It was very clear that Bruce Jones was a marine engineer that clearly understood the power of fusing the two concepts together in ensuring the success of a commercial venture.

Aquanaut and French Architect Jacques Rougerie has designed some of the world's most magnificent undersea structures that fuse both striking, even emotional, form with a daring sense of functionality.

And yet, an extraordinary form does not come easily – or for free. Even if a team possesses a remarkably creative mind, a gracefully appointed structure invariably costs more than the tanks and boxes approach. It increases the project's engineering complexity and time, which in turn increases the project's cost. There is a real pressure on the design team to add tanks and barges – proven and simple assets – to ease up on the cost and time pressures. How a simple cylinder will respond to pressure is well known. A cylinder connected to a cube is also a simple calculation. But when the marine architect increases the complexity to polymorphic shapes, the dynamic and static unknowns begin to creep in, the complexity of the structural analysis increases, the time line begins to stretch and the cost goes up. Most small teams do not even have that kind of design expertise onboard nor can they afford it. Hence, the small team efforts will probably always remain simple.

What will the new undersea architecture ultimately look like? The appearance of the NextGen habitats will invariably be linked to the world in which they are placed. Regardless of the function – whether a submarine, surface vessel or undersea habitat – shapes and forms undersea will be related to the physical character of the underwater environment. For example, it is no accident that the first submarines were shaped like tubes. A tube under uniform compression will hold its shape until it begins to deform under too much pressure. A sphere will also hold its shape under uniform compression. A six sided box will not. The pressure will act to deform its sides and tear its joints apart. In comparison, a tube and a sphere have no joints!

In one of my habitat designs, the basic shape was cubic polymorphic. The walls were made of 0.375 inch plate steel. Its relative interior pressure at the floor was zero but at the top it was 2.7 psi. Even with 0.375 inch plate steel under the modest pressure of 2.7 psi, the walls deformed slightly at the top of the structure. It was nowhere near any danger point and only one other individual noticed the slight bulge – the station's chief engineer. But it clearly and visibly illustrated that shapes and forms undersea respond directly to the environment into which they are placed.

Long before the habitat entered the water for the first time, I had anticipated this deflection but I did not have the budget to determine its extent on the habitat skin as well as the wide windows in advance. I designed and installed the windows' cross bracings to counter that deflection as well as for protection against impact. At the center of each brace, Habitat Engineer Joseph M. Bishop and I installed hard rubber pads that pressed against the midpoint of the window. As far as my eye could determine, the braces worked beautifully and there was no visible deflection at all on any of the windows.

As a former US Naval Officer, I served on many ships of all types and ages. One thing I noticed was that on the older ships, at or below the water line, sometimes the hull plates were slightly deformed inward between the structural members. One engineering officer explained to me that over time and weather events, those plates were deformed by the constant and often changing pressure of the ocean pushing against them. Considering the plates were made of steel considerably thicker than 0.375 inch plate, the sheer power of the ocean's effect on the thick, steel plate was unforgettable!

I have imagined multiple habitat designs and many of them are neither spherical nor tubular. And yet the marine environment is not at all interested in my structural creativity. The ocean is going to do what it has always done – exert pressure and force at its will. My primary task as a habitat designer is not to be creative, but to be creative only after I have considered the power of the undersea world.

In my Aaron Seven novel titled *Quantum Storms*, I created an undersea colony called *Pacifica* made primarily of a single massive structure suspended above the Hancock Seamount in the North Pacific Ocean. As a writer I am not constrained by cost and other realities, so I had free reign on what the massive habitat would look like. When I had completed the design, it was a single sphere attached sphinx-like to a pair of rectangular shaped structures that ran parallel beneath it. Looking at that design, I see a structure that, once again, is somewhat reliant on the basic shapes but one that is perhaps inching its way to a more pleasing aesthetic.

As I consider the model of *Pacifica*, I am reminded that undersea architecture will proceed in such fashion – the simple inching toward the more complex. If we look at the earliest photos of San Francisco, we do not see elegant skyscrapers. Instead we see box-like wooden shacks, all shaped nearly identically, astride muddy streets filled with horse drawn rectangular wooden wagons. In the undersea kingdom that will begin soon, things will be the same. No one is going to begin by building elegant undersea structures right away. But, just like San Francisco, the shapes and the forms will eventually change from the simple to the magnificent.

If we allow ourselves to imagine the future when Aquatica is fully populated, when there are magnificent undersea cities

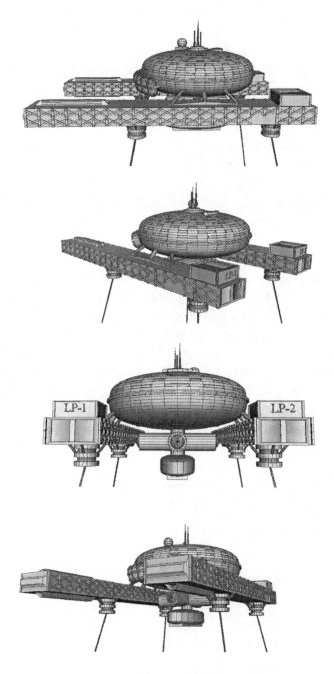

VIEWS OF PACIFICA
FROM *QUANTUM STORMS*

everywhere, what will the individual structures look like?   I believe, if I am allowed such free running imagination, that they will be conformed to the environmental realities simply because they must! The sea is most powerful and most unforgiving of even the slightest error in action or judgment. Hence the structures will be designed around that unforgiving nature. Undersea habitat designers must think first in terms of the environmental realities, conforming their creativity to that nature. However, it is essential to note that a conformed creativity is not at all the same thing as no creativity!

The structures will be shaped to most effectively deflect forces of current and surge. The larger structures will take shapes that allow for the movement of complex surges made up of variable vectors so that a passing surge will be permitted to twist and pull the structures along their entire length without danger of breaching them. If we back off and close our eyes and imagine what such a structure would look like, they must, by definition, be as streamlined and smooth as a jetliner – able to handle pressure effects while at the same time able to hydro-dynamically deflect passing current and surges. They will be shaped to undergo pressure changes and static pressure without significant deformation. After those considerations, the designer is free to release their maximum creativity and design a whole new generation of human architecture for a new and very different world undersea.

Yet even the most remarkable colonies are useless unless the Aquaticans are able to access them!

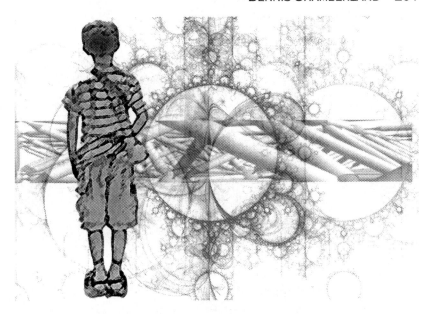

COMMUTING

"Two roads diverged in a wood, and I – I took the one less traveled by, and that has made all the difference."
Robert Frost

 f all the aspects of living permanently in the sea, there is perhaps no other more significant initiative than commuting.

How do we get there, and then how do we move about?

The answer to this question is the single most important key to the expansion of the undersea world to permanent human occupation. Why is it so important? Because if people have any thought that they are in some way trapped underwater for whatever reason for just about any period of time, there will be little interest in living in an undersea colony, and I will definitely be included in that number!

We are a free people who travel freely and often, and no one likes the feeling of getting stuck. I am a healthy man who wishes to see my six children and their children and I have a lot of traveling left to do in my life. And whether my address is in Cocoa, Florida, Stonebrooke, Tennessee, or the Atlantica Undersea Colony, my wishes do not change. I want to come and go whenever I please, just as I have always done.

It is most unfortunate that in the past history of undersea habitation the notion was hatched and communicated widely that a long and dangerous decompression period was required to come and go after living for even a single day in an undersea habitat. While that was true of the first habitats based on their mission profiles, it is but one of many scenarios that is possible for living undersea.

Indeed, experience has aptly demonstrated that such a scenario is undesirable and most impractical.   Therefore, for permanent habitation, undersea habitats will be designed around a no decompression requirement to effectively and efficiently get the Aquatican up and down. When creating the undersea colony, this will become the most important part of the design process.

The philosophy of commuting is critical because it represents *freedom of access.*

When aviation first began, aviators sat in open cockpits and dressed warmly. It was still often a miserable experience – nearly always cold and sometimes very wet. Later, closed cockpits were invented, but it was still cold and when the aviator flew high, they were forced to don oxygen masks. All in all, it was still miserable. For the most part, only aviation professionals left the ground and it was often dangerous – never an activity you would take the kids on for a weekend outing.

Obviously, when the heated and pressurized cockpit was invented, things changed straightaway.   From that point on, prospects for aviation immediately reversed themselves.   Now, tens of millions of people fly each year and hardly any one of them can pilot a jetliner. No one gets wet; no one gets cold; lunch and a movie are served at 40,000 feet. The first pilots would be totally astonished – and some perhaps annoyed.   Now instead of a handful of stalwart, courageous 'aeronauts', there are tens of millions of them – and some are even infants!

In the near future, Aquaticans will commute just as freely as their friends and family on dry land. Depending on the distance the colony is located from land, they will take various forms of transport to and from the undersea colony. Cousteau envisioned this when he laid out his plan for Conshelf II. In his footprint, he staged his *Starfish House* – the main habitat – adjacent to the underwater hangar for his submersible, the *Diving Saucer*. Unfortunately, his design did not permit the *Diving Saucer* to dock directly with the *Starfish House* so that the transit to and from the surface could have been dry and, because the divers were saturated, they could not take the saucer to the surface. But he was on the right track.

COMMUTING FROM LAND BASES BY HYDROFOIL
CONCEPT AND GRAPHIC © BY GUILLERMO SUREDA BURGOS

A key enabling philosophy for undersea colonies is the development of a system that allows a transfer to and from the surface that is dry, safe and non-invasive. It can be a submersible, bell, sea elevator or a hybrid of these technologies.

In the former Walt Disney World Living Seas pavilion, visitors were able to visit the *Sea Base Alpha* pavilion by 'descending in an elevator'. Such a system – although nothing more than a necessary part of the staged presentation to move large numbers of people in and out – is still a brilliant picture of the ease of movement that average people will expect for relatively effortless access.   Admittedly, the Living Seas elevator was probably a little on the 'too simple' side, but it does clearly illustrate a system of dry, safe and non-invasive commuting into and out of an undersea colony.

If you recall the system that has been discussed throughout this book, it is one where the undersea habitat is a structure where the main living areas are pressurized from one to no more than 1.6 atmospheres. When the designer links the transportation and commuting scenario to this system it becomes evident that if the Aquatican is pressurized to 1.6 ATM, he or she may travel freely to the surface. However, they will still be restricted from flying for 24 hours because of the danger of bends in a lower-than one atmosphere pressure in the aircraft. Hence, while the Aquatican is unhindered in their movement between the undersea colony and the surface, they may find a 24 hour constraint on flying in a fixed wing aircraft or a helicopter if the colony is pressurized to 1.6 ATM.   Such considerations may well limit the colony to a one atmosphere pressure or it may require the design of a one atmosphere section for those who plan to fly the following day. It is even conceivable that individual family compartments may have the capacity to alter their pressures from 1.6 to 1.0 atmosphere.

Today, over 90% of all commuting to and from functioning habitats is by free diving. In most cases, it is by slipping in and out of a moonpool in a 1.6 ATM ambient pressure habitat and swimming directly to the surface.

This all makes various assumptions about the aquanaut and the system: they are trained and certified to dive (which also assumes they can swim and are not afraid of the water); that no decompression is required; that they will not fly within 24 hours if they have been saturated at 1.6 ATM; and that the weather conditions at the surface are permissible for swimming up and boarding a watercraft. While these assumptions may sound overly simplistic to a trained and experienced aquanaut, to the average non-diver they are not so simplistic at all! Indeed, they are limiting and, ultimately, disqualifying. In fact, most people would never even consider this activity, regardless of how simple it sounds to certified SCUBA divers.

It is crucial that the current community of aquanaut professionals never step over the fine line of being concerned about safety and training to becoming elitist by actually enforcing the disqualifications into system design and a philosophy of, "If I can do this, why should I worry about someone else?"

In my career fields of scientist and bio-engineer, the most frequent ruse by 'professionals' to preserve elitist positioning is the complaint that someone is not 'certified' or 'adequately trained' in their chosen field of endeavor. This did not seem to be a problem with the various tourist astronauts of recent years who showed up to the launch pad with minimal training but a whole lot of cash. Nor did it prevent relatively novice politicians from riding into space.

The new Aquaticans will be people who have but a single qualification – they want to live permanently undersea. As one aquanaut put it, "Do you have a heartbeat?" If the answer is 'yes', then you're qualified! At this point I can hear the collective groaning of the Gentlemen Adventurers Club members. And I can assure you that all their collective harrumphing is not a concern principally over safety issues. The problem is that the good-old-boys club is abut to be raided by the 'women and children'.

But these guys are, fortunately, not many in number. The true professional aquanauts, professional divers and diving community will work to make that dream come true for everyone. In the undersea kingdom, it is not just the wave of the future – *it is the future*!

Likewise, commuting is the most critical aspect of the future of undersea colonization because, in a single step, it eliminates the chief disqualifiers for access. It does not just represent the doorway of humanity into the seas – *it is the doorway*!

If you were to poll the average citizen from nearly any country on earth and ask them if they would be willing to be trained to certification, then put on SCUBA gear and dive down to an undersea colony for a visit, then don SCUBA gear for the trip back to the surface, probably 95% of everyone asked would respectfully decline the offer. But if you could assure them safe and dry access in both directions with no limits to their return, it is very probable that those statistics would nearly reverse themselves. Hence, stating it again:

The philosophy of commuting is critical because it represents *freedom of access.*

For example, I cannot adequately express the level of excitement that was generated when my team received the submarine, *Dan Scott Taylor* II (DST-II). It is the first submarine in history ever designed to deliver humans comfortably and dry in one atmosphere to an undersea colony.   The interest and enthusiasm even exceeded any we ever received for any of our habitats. Why? Because the DST-II represents freedom of access for everyone to a deep, blue water undersea colony, and there is no demand to expose anyone to the open ocean in the process.

Utilizing various approaches of commuting to and from and even around the colony itself, the philosophy invokes the First Principle of Undersea Colonization: "In every exploration system, we must require the systems we build to adapt to the human standard rather than expect the human to adapt to the machine or the environment – and in every design activity we will protect the human as a primary objective."   Let us now examine the various commuting scenarios that will probably become standard systems in an undersea colony.

Getting In and Getting Out

There are at least four forms of engineering solutions for commuting from and to the surface discussed below. The point behind all of these is accessibility. The most damning thing any undersea colony can say about itself is 'restricted'. If it is restricted because of inadequate access, for reasons of conditional departure, or a requirement of intensive training, then the engineering and planning that went into that colony is either primitive, deficient, unimaginative or all of the above.   It is true that this principle sets the past history of undersea habitation right on its head, but it is also quite accurate. To get into an undersea colony, there must be a door. If there is no way in or out, or if the way is unrealistically

THE DAN SCOTT TAYLOR II SUBMARINE

THE DST II IS THE FIRST SUBMARINE EVER DESIGNED SPECIFICALLY TO SURVEY POTENTIAL UNDERSEA LOCATIONS FOR A PERMANENT UNDERSEA COLONY, THEN PREPARE THE SITE FOR THE COLONY STRUCTURES AND HABITATS. THE 48 FOOT DSTII IS A LEAGUE OF THE NEW WORLDS ASSET.

restrictive, then, as Jacques Cousteau and the United States Navy found out, there will never be an undersea colony to discuss.

Here are four ways that will allow access to just about anyone who wants to visit an undersea colony:

SCUBA Diving: Some hardy souls will always want to enter the colony by diving. They will want to leap off the S3 (Surface Support System) platform and head down into the briny deep, armed only with their courage, wits and stainless steel dive knife between their teeth. Fine. We'll leave the lights on for them and an open moonpool. Dry towels and hot showers will be provided at the end of the adventure.

Submarine: Perhaps the most exciting way to arrive in any undersea colony will be by submarine. The DST II will one day soon be docking with an undersea colony on a regular basis.

These submarines are designed for the pick up of passengers, either at a land based pier or at a surface S3 platform, for transfer of personnel to the colony. Commuters enter and exit the submarine dry and they return to the surface dry. In one plan, the sub is designed to mate with the docking station at the bottom of the colony and open its hatch at the same pressure of the habitat. In another less sophisticated design scenario, the submarine will 'blow down' to depth (increasing the pressure to the pressure of the moonpool) after a hard dock at the open moonpool, and unload the passengers who will then enter the colony though an airlock. This can be an unpleasant and stressful maneuver, so it is far less attractive than a sealed docking at habitat pressure.

The submarine provides an unparalleled level of excitement because the aquanaut is taken down in a shirtsleeve atmosphere, untroubled by waves, surge, current, water temperature, personal stress or any other unpleasant environmental conditions. The submarine may orbit the colony before docking to give the arriving aquanauts a clear view of the extent and setup of the entire settlement from a distance – a view not afforded in any other way. And while there, the aquanauts are then able to fully enjoy their stay, knowing that their return voyage will be just as pleasant.

Bottom Staged Tethered Diving Bell: Diving bells have been used for centuries (even by Alexander the Great!) to transport people from the surface to the undersea world. But the function of the diving bell used in the undersea colony is reversed from the traditional diving bell method. Instead of being lifted off of a surface vessel, it is mounted to the colony's frame (or seafloor) and winched up to the surface S3 station where it will pick up passengers. Once loaded, the bell is then winched back down to the lower level of the colony. The entire base of the diving bell can then be mechanically turned so that it is lined up with its docking collar. It will then be raised and mated with the habitat just like a submarine.

Again, these bells allow for transfer of personnel dry and comfortably in shirtsleeve environments in both directions with no stress caused by the surroundings or for expectations of their performance.

Fixed Bottom Staged Elevator: The difference between this transfer method and the diving bell, is that the elevator rises and falls enclosed in a fixed caged structure. The diving bell is allowed to sway in the current and swell as it moves in both directions. But the elevator is caged in a fixed structure that ensures its smooth travel.

The fixed elevator will be large enough to feature seats so that the aquanauts will have a comfortable ride. At the colony, the cage will swivel away and the elevator will rotate up into the moonpool and dock just like the diving bell and the submarine.

When not in use, the cage rotates or telescopes down to the colony so that it minimizes becoming a navigational hazard, its exposure to damage, and its surface connectivity in accordance with the Second Principle of Undersea Colonization: "Every colony system will strive from the outset to minimize surface connectivity. Every possible surface connection will be re-deployable back to the colony if the need should arise."

SUBMARINES FOR COMMUTING IN AND AROUND THE COLONIES
CONCEPT AND GRAPHIC © BY GUILLERMO SUREDA BURGOS

Commuting in and Around the Colony

Local Submarines: Small one to four man local submarines designed for dry transfer between parts of the colony that are not connected by tunnels or passageways will be available. While it is true that in an effective undersea colony design these units will be rare, there are reasons to have parts of the colony that are not connected.

Further, there will be new colony units that may be delivered if not built in place and they will have to be positioned and checked out before the complex process of connection begins. And there will be distant habitats, work modules and research units that are staged much too far from the colony to be connected by tunnels or passageways. All of these scenarios may require the use of a local submarine for transportation of aquanauts and equipment.

Hydrosleds: Hydrosleds, as I envision them, will come in two varieties. One will be a small manned 'wet' platform onto which equipment can be strapped. In the front of the sled is an open position where an aquanaut lies or sits to control the sled's propulsion and guidance. The sled is loaded up and dropped into the open moonpool where it is then boarded by an aquanaut who then turns on the motor and drives it underwater to its destination, delivering the cargo. The second variety will be an unmanned sled pulled behind a local submarine or ones that are remotely piloted from the Command Center.

Personal Undersea Scooters: These small, hand held propulsion units have been used for decades. They consist of a small propeller driven jet that pulls an aquanaut behind it and is directed by onboard controls. They will still have a place in the future of undersea colonies.

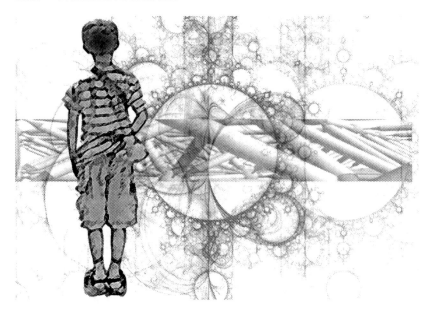

THE S3 — SURFACE SUPPORT SYSTEM

*"Those who say it cannot be done should not interfere with
those of us who are doing it."*
S. Hickman

I t is very possible that one day undersea colonies will
have no 'visible' surface connections at all. After all,
the Second Principle of Undersea Colonization states
unambiguously: "Every colony system will strive from the outset
to minimize surface connectivity. Every possible surface
connection will be re-deployable back to the colony if the need
should arise." The reason for that principle is not at all one of
stealth or intrigue, but that it is the surface interface that is the
source of so much relative chaotic energy and ultimately is so
difficult to manage – as the earliest manned undersea experiments
discovered.

At the initial phases of colony design, a surface interface
system will almost certainly be required. In a preliminary
undersea colony, the community cannot periodically surface like a

submarine to meet its interface requirements. But the colony will maintain some kind of permanent or semi-permanent surface presence called the 'Surface Support System' or S3.

The purpose of the S3 is to provide services that can initially only be obtained at the surface and not undersea. Here is a brief list of such services: power from solar, wind or ocean swell; an air source; communications; docking with surface vessels for personnel transfer; a platform for receiving helicopter aircraft; and all surface logistics connections to receive supplies.

Every one of my habitat designs features S3 systems. Shown above is a photo of the S3 buoy taken from inside one habitat looking up toward the surface. This unit was a relatively small S3 serving a three man habitat. It provided the surface connections for communications, hazard and warning lights and surface weather information.

In the case of several habitat programs, the S3s were large surface support ships. The liability with relying upon a surface support vessel is that if there is a problem with the ship or bad weather is approaching, the mission must be terminated in order for the ship to move to safety. Such events have occurred in the past including during the Conshelf and SEALAB missions. *Aquarius* eventually switched its 100x50 foot catamaran style barge for a semi-autonomous buoy.

In a permanent undersea colony, the S3 platform will need to be a ballasted structure so that it can be safely lowered to the level of the colony if the need arises, wait out the passage of the problem on the surface, and then resurfaced as necessary when the interface conditions allow. It can be made of steel or new polymer materials that never rust or corrode. Or, it can be designed as a wholly inflatable structure. Again, one of its primary requirements is the capacity to submerge and re-deploy back to surface duties. To rely on a surface support system that will end the mission if it must be taken out of service is certainly not a satisfactory design for a permanent undersea colony and can never even be considered! Because of the vagaries of surface conditions, colony assets must be designed to periodically do without the S3 services for critical periods of time.

Depending on the size of the colony and the requirements and design of the S3, the surface system may well be manned from day to day or even 24/7. If the colony receives daily personnel and supplies, then it is probable that citizens of the colony would report to work on the S3 platform each day and return to the colony at the end of their shift.

The S3 is capable of many important surface interfaces. These interfaces include:

Communications: Except for military ELF frequencies, undersea communications use transmission and receiving frequencies which can only be sustained in the air. The S3 will support these antenna systems out of the water and extend them to a nominal height above the surface. Such communications allow primarily for colony-to-satellite transmissions and reception, but VHF frequencies allow for communication with nearby line-of-sight shore services or surface vessels approaching the colony.

The S3 will also supply satellite based Internet communications that will double as private telephone connections, news outlets and email sources for the colonists so that regardless of where the colony is located, its citizens will never be out of touch with the rest of our global community.

Antennas for reception of weather information and data will also be an important segment of the S3 array as well as radar for incoming navigation aids and collision avoidance systems.

<u>Air</u>: Depending on the early colony design characteristics, oxygen may be supplied using surface air as the carrier gas. If the colony's compressors are large enough and energy is plentiful, the S3 may supply a steady stream of fresh surface air to the colony continuously. At the least, the surface connection will allow the colony's pressurized storage and air buffer tanks to be refilled on a periodic basis.

<u>Receiving and Docking Platform</u>: The S3 acts as a floating platform to receive personnel and supplies destined for the colony or from the colony back to shore. These arrive by surface vessels and from helicopter or seaplane. The S3 acts as the docking station for colony submersibles, diving bells and/or elevators, as well.

<u>Power Generation</u>: The undersea colony may utilize *in situ* resources for power generation such as undersea current turbines, surge generators or hydrothermal energy, but much of the colony's power will come from the surface. The S3 acts as a focal point for this power in providing a staging platform for wind generators, solar panels, and swell generators as well as traditional fossil fueled generators.

Wind Generators: Wind power is a highly efficient source of natural power – *if the wind is blowing*! These generators must be placed on the S3 in positions that can be sealed if the S3 is required to submerge, and then automatically redeployed upon surfacing.

Solar: Solar panels are the most problematic power generators for an undersea habitat scenario. They are generally flat but have large surface areas which require that they be supported by a rather large S3. Unfortunately, a large S3 becomes exponentially difficult to maintain in average seas and is also more complicated to submerge. In my book *Quantum Storms*, I solved this dilemma for the ocean colony *Pacifica* by floating the solar panels separately from the S3. When seas became too rough for the panels, I allowed them to slip edge up by de-ballasting one end so that they floated vertically in the storm seas. Like reeds in the water, the forces of the seas could not damage them as easily in this configuration.

Swell Generator: A swell generator uses the ocean's rhythmic rising and falling (swell) to generate power. By placement of equipment, either tethered to the bottom or weighted below the S3, the ocean's mass or even the mass of the S3 platform itself can be effectively used to generate power for the colony below.

Gasoline or Diesel Generators: These old standbys may be old school technology, but they have proven to be the reliable backbone of many undersea habitation schemes in the past and probably will continue to be in the near future.

Caution and Warning Buoy: The depth of the average undersea colony will probably not represent a hazard to most surface navigation. However, there is some measurable degree of risk from surface vessels such as cable layers, trawlers, etc., from falling anchors (in shallower colonies), and from items casually tossed from ships. The S3 acts as a caution and warning buoy signaling ships not to cross over or nearby the area. All umbilicus lines from the S3 to the colony must also be protected from snagging.

The S3 is a vital component to any undersea colony. During the early years, they will be more essential than later when more and more resources are drawn from the undersea environment. But there will always be some form of S3, primarily to receive supplies and personnel from the land based regions, a process commonly referred to as logistics and resupply.

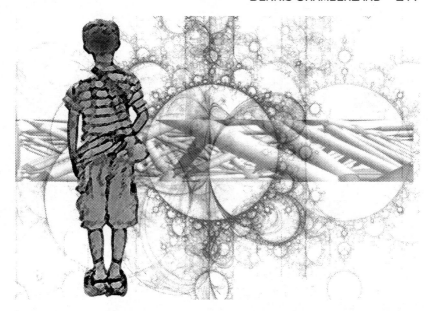

Logistics and Resupply

"Gentlemen, the officer who doesn't know his communications and supply as well as his tactics is totally useless."
Gen. George S. Patton

T he undersea colony is no different than any other community ashore. It will require supplies and services to keep the population alive and functioning at their peak. It will involve every aspect of daily living that shore based communities require; from food to medical supplies to the latest movies. While it is true that the colony will strive for as much autonomy as is practical, most will never totally sever their connections with the outside world any more than any other normally functioning community.

The requirements for undersea colonies will include a wide variety of necessities equal to the needs of any town. Most information can be transmitted electronically, but food, pharmaceuticals, fuel and an early endless list of sundry items will have to be delivered to the colony's S3 platform and then transported underwater.

There are numerous land based assets that will serve the needs of the undersea colony for many decades – such as pharmaceutical corporations and their research campuses located all over the globe. Obviously, no single colony will initially be able to reproduce this invaluable service. Further, specialized hospital assets will probably not be available for many decades in undersea colonies as well as many other resources that have been built up ashore over centuries. The undersea colony, although remote, will still require all these wonderful human services.

All materials and supplies for the colony will be delivered to the S3 platform in the least expensive way; primarily by surface vessel, or the occasional submarine, aircraft or helicopter. Obviously, the further away the colony is from the supply point, the more difficult and costly will be the delivery. Once you reach an extreme distance, most of the colony's supplies will probably only be delivered once per week, or even less frequently, by ship which would be the most economical form of transport, but also the slowest. Deliveries of more sensitive and urgent supplies would come by plane or special fast vessel. The supplies will be logged in, then loaded on a colony submarine or other logistics asset and dispatched to the colony beneath the surface.

In 1991, I created a system for undersea colony resupply I called the 'Logistics Pod', or L-POD. The L-POD is a submersible carrier that comes in many sizes. It can arrive preloaded or be loaded with supplies from the surface at the S3, then closed and sealed. The L-POD is then submerged and a winch-tether pulls it down to the colony's receiving bay  where it is docked, unloaded and returned to the surface for another load. It is created to be a kind of low tech aquatic 'dumb-waiter' to quickly transfer a large quantity of goods up and down the water column. I am shown in this photo testing a small L-POD in a Florida spring.

During our aquanaut missions, our logistics system consisted of suitcase sized, waterproof cases manufactured by Underwater Kinetics. Our resupply process was described in detail in the earlier chapter titled, "One Day In the Life of An Aquanaut". But in NextGen habitats, the systems will have to be larger and carry much more cargo to and from the surface. Further, it will not be a task suited to a 'duty aquanaut'. In habitat missions with four or more aquanauts spending longer periods undersea, new delivery systems, like the L-POD system, which make automated deliveries will be necessary. On some days, the weather or the visibility will prevent the duty aquanaut from making deliveries; therefore an automated system will provide much more efficient and safe transport. As large habitats grow into colonies, it is possible that some daily runs will have to be made with manned submarines while other materials can be carried adequately by L-POD deliveries.

Personnel are also a part of the logistics chain. Personnel transportation to and from the colony may be accomplished by any of the modes addressed in the chapter, "Commuting". In early colonies, emergency transport to better equipped hospitals or trauma centers will probably be accomplished by helicopter or fast boat, depending on the distance to shore.

Finally, it is very tempting to think of logistics and supply chains as hopelessly boring. That is exactly why I opened this chapter with the frank remark by General George Patton. He found out the hard way that you can be the best tactician in history and still get your pants beaten off by the world's worst military mind if he has a good supply chain and you do not.

On one Internet blog titled, "Who Said Supply Chains are Boring?" the author posted this comment:

"When I tell my friends my profession is supply chain management I immediately see glassy eyes followed by a yawn or two. Little do they realize that every time they purchase a CD or a carton of milk they are completing the end of the supply chain cycle. Without supply chains consumers would be stuck using products that they had to make or grow with their own two hands."

One certain way to ensure defeat of undersea colonization is the failure of maintaining a well designed, well planned, well managed and well executed supply chain system – one that links the surface world with the three dimensional world of Aquatica.

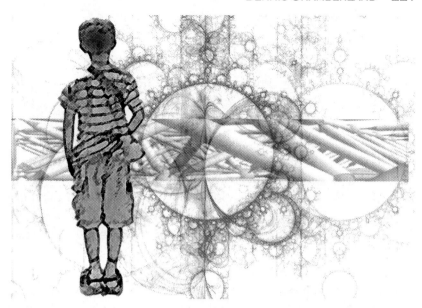

The Three Dimensional Kingdom

"Keep thou thy dreams – the tissue of all wings.
It is woven first from them;
from dreams are made the precious and imperishable things,
whose loveliness lives on, and does not fade."
Virna Sheard

A s land dwellers, it is difficult to think outside the boxes we have been confined to on land. Here on land we are restricted by two dimensions and gravity. Hence our thinking is often two dimensionally oriented. For example, I have heard and read on occasion that the oceans are quite limited for human colonization because of the ocean depths. Indeed, the average ocean depth is 12,200 feet and the majority of the abyssal plains that cover most of the globe average a depth of 18,000 feet, far below any hope of human occupation given current technology. Therefore, the saying goes, human colonization of the seas is limited to the continental shelves, of which fully half are below 'reasonable settlement depths'. Such an idea is even implicit in Jacques Cousteau's *Conshelf* expeditions.[10]

I try hard not to smile when I hear this. It is a clear indication that many times humans are hung up on their two dimensional view of the world. You see, the ocean is not a two dimensional place! In the ocean we are not bound by gravity.[11] And we are certainly not limited to placing our colonies on the seafloor!

When I first began to develop undersea colonies in my mind as a teen and built small insect colonies and placed them on the bottom of my backyard stream, I was bound by the two dimensional idea that the colonies *must* sit on the bottom. As the years passed, I began exclusively planning for seafloor bound settlements. In fact, the idea was so pervasive in my mind that as I considered writing this book in the early 1990's, I originally titled it *Seafloor Colonies*. I even considered naming a colony I had planned the *Atlantis Seafloor Colony*. I certainly have no room for smugness in this area as my own past was firmly caught up in a two dimensional view of the ocean.

My thoughts were changed in this regard by the influence of Gerard K. O'Neill his classic work, *The High Frontier*, who postulated that human colonies in space have no need to be gravitationally bound to planetary or other surfaces. He reasoned that colonies could be constructed by humans such that they created their own surfaces! As I considered this I came to realize

---

[10] Of course, Cousteau was planning for his undersea presence as one pressurized to depth, so that his colony would have required mixed gasses which, at the time, were certainly limited to the shallower confines of the continental shelves. It is no wonder that Cousteau eventually found this idea 'impractical'!

[11] It should be evident that there is as much an expression of 'gravity' undersea as anywhere else on the earth, or even in earth orbit! Because of factors of buoyancy, humans or their colonies can be made 'weightless' with respect to their surroundings and they can 'float' in the undersea medium. In earth orbit, humans and their space stations 'free-fall' around the earth leaving the impression of zero gravity. Yet in both cases, undersea and in orbit, the gravitational field of the earth is still fully present and wielding its force.

the same is true undersea. Since the first elements of future colonies will probably be pre-constructed and then brought to the undersea site, there is no reason why they have to be bound to the seafloor at all. They can, with relative ease, be tethered to the seafloor and suspended at whatever depth the designers choose. Further, as our technology develops, it is also probable that the suspensions can even be done away with and the colonies will 'hover' at whatever depth or position they choose.

In one very powerful experience, I was diving in water so deep the bottom was far beyond what I could see. As I hung there in the void and looked around me, I observed the ocean stretching out before me in all directions, blue and vast. Below me the abyss turned to a deep cobalt blue. As I hovered there, I realized this was much like space, hanging weightless and suspended, yet with full freedom of movement. I could swim up, down or in any direction I pleased. Unlike space, I was not bound in my movements by the complex calculus of orbital mechanics.

Nor was I bound by the bottom that was so far away I could not even see it. Here I could build a colony at any depth I pleased. The seafloor lost its importance. Indeed, use of the seafloor became supremely limiting as the whole of the ocean emerged to me at that moment as an immense three dimensional kingdom that at once was opened to humanity's grasp. No longer were we bound by the uncertainties, permanent darkness and vagaries of the abyssal depths. The vast oceans in their entirety, and not just the continental fringes, could now be a part of the human domain. And this domain was not limited by two dimensions or a very distant seafloor.

The three dimensional concept is like a land dweller deciding whether he wants to live bolted firmly to the ground or hover at any altitude above the ground he chooses. Except for perhaps a balloon mounted structure, the reality of such a thought is not possible. It is not a thought construct that has ever entered the human model of habitation. Hence, it is not surprising that the first discussions of undersea habitation mostly centered on seafloor structures.

As a designer of undersea habitats and colonies, it is an interesting exercise to decide which choice is best for placement of the settlement. Obviously, the seafloor situation is somewhat more straightforward. The suspended colony is additionally complex but grants a wide degree of freedoms that are unavailable to fixed colonies.

In the seafloor colony, the idea is to find a level surface, deploy the colony structures in such a way that they are very negatively buoyant and then provide scour protection around their bases. In the advanced planning for the colony, the site would have to be thoroughly data logged for prevailing conditions, and the colony structures would be oriented for the least impact from prevailing currents, surge and long period swell effects. The concerns are that the colony structures not be scoured out of place, lifted, pushed around or disassembled by these undersea forces. It is a rather simple approach that offers the benefit of a stable bottom on which to fix the colony by means of sheer mass, orientation and intelligent design based on known threats.

The suspended colony approach and design is quite different. As I stated previously, in my book *Quantum Storms* I developed the scenario of a colony named *Pacifica* suspended by cables over the top of a seamount. The colony hung in the ocean void 200 feet down from the surface and six hundred feet above the top of the seamount. As an exercise, let us use this colony as a point of discussion.

As we plan to place *Pacifica* over the top of the Hancock Seamount, we are aware that the idea is reversed from a seafloor colony. Instead of purposefully seeking very negative buoyancy, we are seeking positive buoyancy. The positive buoyancy will serve to keep the colony suspended instead of dropping to the bottom. Careful buoyancy management is a significant aspect of this design.

The degree of positive buoyancy is a critical part of this paradigm. If it is too near neutral, then any change in the colony's mass could cause it to crash to the bottom, catastrophically destroying it. For example, in the event of a loss of a compartment or two due to flooding, the whole structure would have to be positive enough to maintain net positive buoyancy or this off-nominal change in buoyancy would cause the entire colony to sink. The solutions to this hazard would either be to maintain very high positive buoyancy all the time or have emergency buoyancy tanks available to blow down in the event of a crisis.

Maintaining high positive buoyancy may not be the best solution. The higher the positive buoyancy, the higher the tension on the anchor cables. In the event of a long period swell or storm, this will cause a shallower colony at a depth equal to or less than 200 feet (or sometimes even lower) to 'feel' the swell. In these cases the structure will be moved about by the undersea swell. The degree of motion is, of course, dependent on the wave characteristics on the surface, as we discussed previously.

In the case of a long period swell, the entire colony will follow the movement of the swell in a circular path, regardless of its size or mass. This could cause the colony to sink a few inches or feet, removing the tension on the cables, then rise with the water's motion, effectively 'slamming' the cables with the entire mass of the colony as it rises with each cycle. Obviously, it would not take but a very few cycles before this motion parted one or more of the cables. This would result in the colony rising on one end or perhaps even breaking free and rising to the surface – a very bad day for everyone.

To mediate this possibility, it would be necessary to install very large spring loaded shock absorbers specifically designed to dampen any impulse loading of the cables. Because the worst case scenario would be designed into the colony, these shock absorbers should reduce the storm and swell periods to relatively uneventful occurrences, but ones that would certainly be noticed by the colonists. Another mediation process would be to lower the colony by winching in on the seafloor mounted cables. In this way, it would definitely be possible to lessen the effects and buffer them to more comfortable levels.

In the suspended seafloor colony, it would be advantageous to mount every part of the colony to a single platform except for perhaps a lifeboat station. In that fashion, there would only have to be a single colony buoyancy control system and set of suspension cables. The platform would become an artificial seafloor! To this platform, every habitat could be affixed as well as all connections between them.

One chief advantage of the suspended colony is control of the three dimensional position in the sea, something totally lacking in a seafloor mounted colony. The other advantage is being able to custom design the artificial seafloor to meet the needs of a small colony that has the capacity to grow outward from its original platform to one as large as is required.

Eventually, technology will advance to the point that the cables attached to the ocean floor are no longer required. In this design scenario, the colony will float in the ocean void, maintaining its position with fine buoyancy controls and thrusters, and sustain its preset depth by commands from the control center. If it is in a current, it may send down a single anchor so as not to waste precious energy fighting against the stream to secure its permanent position. But the anchor's purpose is not the same as in a suspended colony – it is only to hold the colony against the stream. The colony controls its depth strictly by an automated buoyancy control. If the colony does not care about its position, it may choose to float in the undersea currents to wherever they may take it, circling perpetually in the hemispheric gyres in a continuous undersea orbit at whatever depth it chooses.

This kind of system is most sensitive to changes in not only mass lost and gained, but also to changes in the water temperature that would affect the density of interior voids. It also has volume and many other buoyancy considerations that ultimately affect the position of the colony in the water. Obviously, such a system will of necessity be very complex and will have to include many automated safeguards against loss of control.

In all these designs, the undersea colony is no longer fixed or bound by the seafloor which becomes somewhat irrelevant and in no way confining or limiting to the expansion of humanity into any part of the three dimensional oceans of the earth.

No matter how colonies are set or suspended in Aquatica, they will all ultimately have a requirement to anchor to the seafloor in some manner.

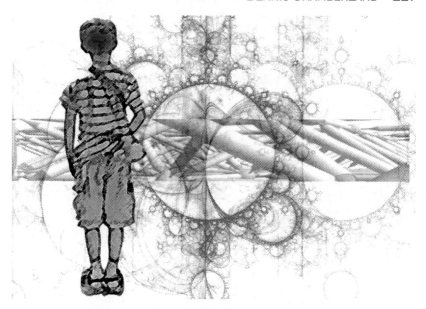

ANCHORING

A ship should not ride on a single anchor,
nor life on a single hope." Epictetus

I magine trying to build a home for your family made out of a thin Mylar sheet filled with helium. The problem would be tying it down so that it would not float away while you were working. Now there is a problem no land bound carpenter has ever had to face. But undersea, it is one of the most difficult problems in any habitation project – the house is always trying to float away!

Not only is the undersea dwelling structure always trying to float away, but it is doing so with enormous force. Unlike the feathery force of a child's helium balloon, the habitat is straining to rise at a strength of 64 pounds of upward force for every cubic foot of air space! At that rate, two average helium balloons could lift a small child off the floor and transport him to Neverland forthwith.

When designing and constructing an undersea habitat or colony, anchoring becomes as important as, or perhaps even more important than, the foundation for a land bound home. In the history of undersea habitats, the first anchoring idea was the simplest one: to wrap chains and weights around the structure until it sank. A few early habitats were thus anchored to the ocean floor, just like the string of a helium balloon being held down by a quarter tied to the end of it. (Having been one of those children enormously interested by such things, I clearly remember playing with such a balloon system for hours – as recently as last week!)

The balloon is trying its best to rise against the weight of the quarter. But the effect of gravity on the coin is greater than the effect of the less dense helium straining to ascend against the greater density of the air. When the system is finely balanced, it is fascinating to lift the string with almost no effort at all and watch the balloon strain against the mass of the quarter, then very slowly settle back down to the floor. The finer the balance, the slower the effect of gravity on the coin. If the balance is indeed fine, the quarter can be shifted from its position on the floor by simply blowing on the balloon with a single, gentle puff!

An undersea habitat behaves exactly the same way. The less dense air bubble in the habitat – some 784 times less dense than water – is always straining to rise against the force of the water that is trying to displace it in a very powerful way. In order to understand this process more clearly, get a five gallon bucket and fill it up to the top with water. Now lift it up. What you are feeling is gravity trying to pull that mass down with a force of 42 pounds. If you dump the water out, put a lid on the bucket and jump into the swimming pool clutching it tightly, the effect would be nearly reversed. It would require a force of 43 pounds to pull that bucket completely underwater! This is an excellent exercise to give you an idea of the forces involved when anchoring a habitat.

If an object is trying to float up toward the surface, it is called, 'positively buoyant'. Likewise, if an object tends to sink, it is called, 'negatively buoyant'. An object that just floats along in the water column, is called, 'neutrally buoyant'.

If the ocean was like your swimming pool, then all the Aquaticans would need to do is to weight their habitats just enough so that they were just slightly negatively buoyant. In the calm swimming pool, that would be fine. But in the ocean, there are many forces affecting the environment that are trying to move the habitat – such as currents and swell actions. These forces are not at all trivial, as we have mentioned in previous chapters.

As we also discussed, there are at least two principal strategies for locating an undersea colony: fixed to the ocean floor or anchored in the void. Both require some attachment to the seafloor, but each approach is vastly different.

In the previous chapter we discussed tethered or suspended colony structures connected to the deep ocean floor by cables and spring-loaded shock absorbers. We can assume that the attached anchors are solid structures, or big dumb-weights fastened with lots of chain and cables. (I realize, that as a technical description, this is somewhat lacking – but, nonetheless, it paints a reasonable picture of what is going on.)

There has been a lot of activity in the past decade or so using ships of all sizes as artificial reef structures. The strategy is to clean the ship of all chemical or petroleum hazards and then sink it so that marine life can find niches in the new, man-made reef. It is a perfect symbiotic relationship between man and the sea: the sea inherits an instant biological niche; the marine life gets a home; and man gets a new reef for fishing and touring! If such ships could be sunk in deep water specifically for anchoring a suspended undersea colony, Aquaticans could save a major cost of colony construction by using the sunken ship as a deep sea anchor. And the ocean would gain the advantages of another new artificial reef!

If the colony is located on the seafloor, there are several anchoring strategies that are useful to keep in mind. As we discussed in 'The Undersea Environment' chapter, it is not as easy as simply blowing the habitat ballast tanks during deployment and settling to the bottom with a slight negative buoyancy, as the poor aquanauts of the *Chernomor* found out. Nor can you count on a significant negative buoyancy and then forget about scouring, as the *Aquarius* found out. The early *Hydrolab* aquanauts also discovered the hard way that just a simple anchor is not at all satisfactory – unless you want to get violently seasick hanging six feet off the bottom! All of these past missions have taught us more than a few valuable lessons about properly fixing our habitats and colonies to the ocean floor.

The first and most simple strategy is an internal air tank arrangement that allows the habitat to carry all of its anchoring mass inside, like the *La Chalupa*. In this scenario, the habitat is to be submerged in a controlled fashion, settled down gently to the ocean floor and, once safely in position, the air tanks be blown of air and filled with water to allow for a significantly negative mass on the seafloor.

However, Dr. Jim Miller recalls an experience with the *La Chalupa* which was sitting on the bottom in 105 feet of water. One day during a mission, the surface was experiencing swells of 10 – 12 feet and their period was such that they were 'touching the bottom'. Even though the *La Chalupa* (whose dry weight is 200,000 pounds) was 20,000 pounds negative, the swells were lifting one corner of the habitat up off the bottom and setting it back down on each pass!

As to what degree of negative ballasting for a habitat is 'safe', there is probably not a good answer because undersea conditions change so radically and each habitat has a different mass and hydrodynamic interaction with the water. As an estimate, given their experience, it is certainly more than the *La Chalupa's* 20,000 pounds positive displacement. I would probably design for *at least* one-for-one negative buoyancy for positive. In other words, if your habitat's displacement is 50,000 pounds, it should be *at least* 50,000 pounds negative on the bottom – but this, too, is just sheer guesswork on my part.

In larger habitat structures, this amount of ballasting is probably either not possible at all or it is prohibitively expensive to design just for the single function of getting the habitat safely down and then negatively buoyant. However, one method that may be possible to achieve a large negative mass is to use *in situ* resources found on the bottom around the colony. Sand, of course, comes to mind immediately. In an early design of the *Challenger Station*, we elected to provide for 'open' tanks where sand from the local seafloor could be pumped into them for achieving a significant negative mass.

But negative ballast can also be achieved in terms of anchoring strategies, not just by providing mass. There are several anchoring strategies for undersea structures that are commonly used today, and some have been for many decades.

The oil and gas industries anchor massive drilling platforms by often using what are called 'suction piles'. Described as simply as possible, a suction pile is a pipe with an open bottom and a closed top. The pile is set vertically on the bottom and the water is pumped out of the top of the closed pipe with a very powerful pump. The hydrostatic pressure of the surrounding ocean literally drives the piling into the bottom! If these piles are added to the base of an undersea colony, it could be towed to its site, submerged, the piles set in place, and then the base is ready for adding incoming structures in this pre-anchored configuration. Further, these pilings allow the base to be set above the surrounding seafloor so that scouring is never a problem!

One other strategy is the driving of piles from surface barges. Such piles are driven for bridges and large buildings. Although principally designed for providing downward support against gravity, they also have a capacity to provide upward resistance against buoyancy as well. Anchoring to these concrete pilings would be a relatively simple engineering and design affair. However, seafloor depth could obviously be a limiting factor for the use of surface driven piles.

Other undersea anchoring strategies have been used. There are Manta Ray®, Platipus® and other commercial anchors that are hydraulically or pneumatically driven into the seafloor. When they are pulled back, they 'deploy' like a toggle bolt and hold against the force pulling against them. These anchors are very common and are even typically found as the cable anchors for large and small utility poles as well as pipelines. Another strategy, if the colony is located near bedrock, is to anchor by drilling and setting anchors into the bedrock itself.

For any of these anchoring strategies, it is essential to plan for the specific strategy well in advance of the actual colony design. In this way, the anchors can be set in place before the arrival of the structures and the base can be specifically designed with the anchor tie-points built in.

Regardless of the anchoring strategy, all habitat designs are energy dependent, and energy is one of the most important considerations in any colony configuration or design.

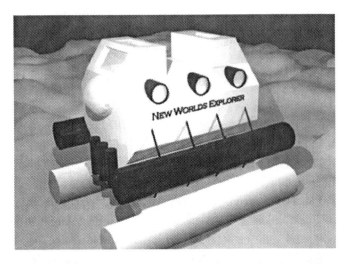

THE NEW WORLDS EXPLORER IS A TWO PERSON — TWO DAY OR "WEEKEND" CAPABLE HABITAT THAT CAN BE LAUNCHED FROM AN ORDINARY BOAT TRAILER FROM MOST BOAT RAMPS. THE NWE HABITAT IS A LEAGUE OF THE NEW WORLDS ASSET.

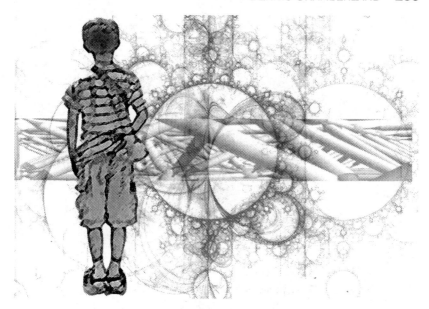

ENERGY

*"Our imagination is the only limit to what we can
hope to have in the future."*
Charles F. Kettering

E nergy is the lifeblood of the undersea colony. If the lifeblood stops flowing, even for a short period of time, the colony cannot survive. Without adequate energy, the undersea colony will fail. In a land based home or city, if the energy stops, the residents can wait it out, albeit with some level of discomfort. Except for cases of medical life support, the residents can wait hours, days or even weeks for the power to be switched back on. But in the undersea colony, by definition, everyone is on life support. If the power goes down, the colony and her citizens are in imminent danger.

As I write these words, I am sitting in my study in my Stonebrooke, Tennessee, hideaway staring out at the forests of the southern Appalachians. We are not connected to the grid here. We utilize a combination of solar power, wind, and a gas powered generator connected to a battery bank and inverter. This marvelous system allows me to type these words on a computer whose energy was derived from many different sources including photons from the thermonuclear reactor of the nearest star, over ninety million miles from my collectors. Here I have the luxury of shutting my personal power system down, as I occasionally do for repairs, improvements or whatever reason I desire. But I also have the comfort of breathing the fresh air of the forest, enjoying the cool breeze and listening to the beautiful sounds of the stream while the Stonebrooke grid is down for whatever reason. And I can do it all in perfect peace, catching a nap between jobs, if I so desire.

If I were in an undersea colony, things would be radically different. I would be sitting in a world of my own making which would include the very air that I breathed, the absorption of my own carbon dioxide and precise humidity and temperature control.

Here in Stonebrooke, all of those issues are taken care of without my help at all. I may adjust the temperature inside, but that is about the extent of it. If the fire goes out at night, I may pull on an extra blanket and that single act will do just fine until morning. All the rest of my life support is supplied by God and the forest without a single thought or action required of me.

In an undersea colony, habitat or submarine, things are not the same. It is the life support system of my own design that keeps me alive. And that LSS is totally dependent upon power. If the power goes down, so does the life support system. And if I lose my life support, I either surface immediately or I will die. The moment the power drops offline in an undersea colony, the clock immediately begins to tick down toward evacuation to the surface. In every habitat I have ever designed, they all included two alarms – a flooding alarm and a loss of power alarm, because both issues are considered emergencies and both require immediate attention.

In an undersea habitat, power is essential for three reasons (in order of importance): life support; life support; and finally, life support. The effect of losing power is immediate on all of the above. While I may sound redundant, I specifically said this to make a point. Permanence in the ocean is directly related to the power that makes the life support system work as designed. If the power is lost, the mission is over.

While it is possible that compressed oxygen or air may be stored in tanks onboard, it must be delivered to those tanks by powered systems. If the power is lost, they can be used as an emergency backup, but very soon (in terms of a permanent system) they are going to run out.

Aside from moving air from the surface to the habitat, the next most important function for energy to perform is absorption of carbon dioxide. The most passive CO_2 systems I am aware of all require energy. These systems are required to mix the air (an energy consumer) into which the humans exhale CO_2, then move that air into the scrubber device (an energy consumer), then blend the scrubbed air back into the air system again (an energy consumer). CO_2 scrubbers all require air to be moved and mixed in any scrubbing application. While some claim that their scrubbers only use the energy required for the 'ventilation system', ventilation is less important than the CO_2 scrubbing task; therefore, in reality, that ventilation arrangement is actually a part of their scrubber system.

During an emergency in sunken or stranded submarines, it is possible to spread the scrubber compound out on flat surfaces and allow passive air circulation to effect scrubbing. But this procedure is recommended only for emergencies and it is dangerous and not very efficient, resulting in high levels of CO_2 building up in the boat. In the end, the CO_2 scrubbing process is always and invariably going to cost the system energy.

After the atmosphere has been provided and CO_2 scrubbed out by using energy, the air must be conditioned by energy. Humidity control requires that the air be passed over cooling coils or, in newer technologies, the air is passed though hydrophilic tubes that absorb the water vapor from the air into a stream of moving liquid. All these processes require energy. Once the humidity has been adjusted to recommended values, the air must either be heated or cooled. (Do I really need to say it?) *Heating and cooling air requires – energy!*

To accomplish these necessary atmospheric duties, the colony incurs its most important energy debt. Depending on all the processes of the colony, it may not be its *greatest* debt, but it is certainly its most *important*. No matter what else happens in the colony, this debt must be paid first.

After the fundamental life support requirements, then the light bill must be paid. Lights throughout the colony are a must, since ambient lighting will not be enough under most conceivable circumstances. As was previously discussed, high efficiency and low heat generating LED lighting may be used to drop the energy requirement. Their usage may be supplemented with other lighting strategies including: hybrid systems that use ambient light; surface light piped down to the colony by light pipe technologies; or light conducing fibers.

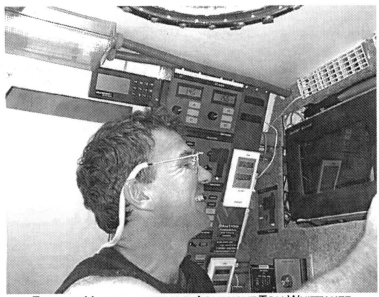

EVEREST MOUNTAINEER AND AQUANAUT TOM WHITTAKER
OPERATING THE MICROWAVE FOR DINNER

Cooking is also an undersea colony energy consumer. It is difficult to imagine that any flames will be utilized in a cooking process in an undersea colony. Hence, all cooking will either be accomplished with an electrical resistance device (such as an electric stove) or a microwave oven. While it is possible to take the cook-less, prepackaged food approach, the crew will almost certainly mutiny after the third main course of beanie-weenies (or you may be able to get by with five main course meals of freeze-dried, pre-packaged military MRE's). You may save energy on one end, but you will pay with an insurrection on the other! Future established colonies will most certainly produce and harvest the majority, if not all, of their food requirements, and these raw ingredients will require energy to be prepared for consumption.

Eventually, even the most stalwart of undersea colonists will need a hot shower. Although the undersea colony will almost certainly not permit their citizens to squander hot, fresh water like an average land based home, it will still require some amount of energy to heat the water delivered to every colonist's shower. In a land based home, this energy requirement is significant, but can be more carefully managed with reasonable and intelligent resource conservation.

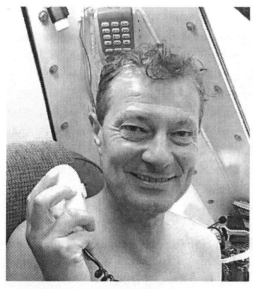

FRENCH ARCHITECT AND AQUANAUT
JACQUES ROUGERIE
AT HABITAT COMMUNICATIONS CONSOLE

Communications in the colony and outside is also an energy consumer. Even though it may not be a significant one, it is still an essential service and it, like all other processes, will require an expenditure of energy which must be considered.

Somewhere near the end of the power requirement scale is colony entertainment. Powering entertainment devices such as televisions, satellite radios, PCs, game and movie systems, and various kinds of power transformers for each device will cost the colony.

As we examine the long list, we must ask ourselves a key question: if there is such an energy requirement for each colony, then where does all this energy come from?

In some habitat systems and habitat facilities being planned for undersea resorts, the power will come from a shore based generation facility. The power is simply routed off the land based grid and run down to the habitat. This system is as reliable as the electrical grid to which it is routed, which means it is acceptable on most days. The only problem with this plan is that sometimes parts of the power grid go down for longer than the undersea colony can tolerate before it would have to be evacuated. It is possible to link to a shore side emergency generating system if that is the case. When a shore based grid is available and joined with an emergency generator, this system is certainly ideal if the colony can pay the bill.

Yet, realistically, these grid based power systems are probably not going to be very common. Just the idea of running a power cable even twenty miles offshore would be prohibitively expensive, even before the first switch was thrown. Yet, any such enterprise is certainly based on its profit scheme, and perhaps such a scheme is feasible relatively near shore.

SEALAB II used a surface support ship to generate most of its power. But we have discussed such a plan and concluded that because any ship's consistent ability to remain on station under any and all conditions is in significant question, this arrangement is probably insufficient, as has been amply demonstrated on several previous undersea missions. Certainly such an array is totally unsuitable for any permanent undersea colony.

Hence, the requirement is for a power system mounted on surface support systems of the S3 variety – one that is capable of remaining on the surface and generating power for long periods, and then has the capacity to button up and descend to the undersea colony and wait out passing unsuitable surface conditions such as storm and high waves. This kind of power system will require that the colony switch to a power restrictive mode and utilize power generation systems limited to battery storage or undersea modes of generation while the surface is abandoned to its hours or days of tempest. Finally, when surface conditions warrant, the S3 can be redeployed and the surface power collection again energized.

Once again, this plan invokes the Second Principle of Undersea Colonization: "Every colony system will strive from the outset to minimize surface connectivity. Every possible surface connection will be re-deployable back to the colony if the need should arise." Taking this philosophy to its limit, it is desirable to attempt to build power generating capacities below the surface interface and its unpredictable and relatively unreliable mood swings. Hence, whenever possible, current power generators and turbines as well as swell generators should be focused on in order to minimize the surface connections and take advantage of the considerably more stable depths. For large, future colonies the question of nuclear power must be considered.

In times of storms and long period swells, if systems are well designed, subsurface power creation may actually exceed surface power generation since the flow of energy through the colony is at its greatest. While these times also challenge the design and implementation of structural and component integrity and the passage of structural components through energy fronts, they also offer an unparalleled opportunity to capture subsurface energy!

In major ocean currents, the flow of energy is astonishing. For example, in the Gulf Stream, a virtual river of ocean water 500 times the flow of the Amazon,[a] passes the eastern coast of the United States each second. Within this vast flowing channel of water, there is more energy available than in all the hydroelectric power generation stations on earth. Hence, any undersea colony situated within this flowing mass of water will certainly be able to capture all the power it requires. While the Gulf Stream is the largest such ocean current, it is still only one of many in the world.

The caveat is, obviously, having a power generator system that is designed for that flow of water. Research has been done on this question and generators have been designed for just that purpose.[b] These generators are very large because they were sized for connecting to shore based power stations, but they can relatively easily be scaled down to colony sized requirements. With such a system, there would not be any power connection or requirement from the surface or from any shore based grid.

New fuel cell technologies will also enable undersea colonies to generate their own fuel from the hydrogen they generate from seawater hydrolysis. Such power generators will literally fuel themselves from *in situ* hydrogen.

Yet, in any human based power system, there will come a time when the lights go out. The colony power system must be designed to handle unexpected power outages for whatever reason. In these systems, the power demand can be shunted over to emergency power circuits powered by a very large battery bank that is kept in reserve just for this purpose. These battery banks can be so robust that they can easily be designed to handle all emergency power needs, including life support, for days or even more than a week. Obviously, during that time it will not be business as usual in the undersea colony – much like during hurricane, tornado or ice storm outages we are familiar with – but the colony will be able to methodically execute any emergency plan or procedure until the problem is repaired.

As we have discussed, power can be an *in situ* resource, drawn from the environment around the colony. But another *in situ* resource that the colony is literally floating in is water. The question is, how do we then go about drinking it?

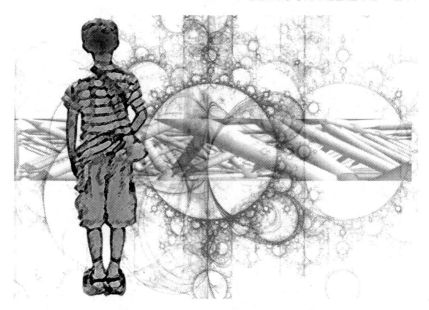

FRESH WATER

"Water, water everywhere and not a drop to drink."
From The Rime of the Ancient Mariner
by Samuel Taylor Coleridge

It has always seemed more than ironic to me that you may live at the bottom of the ocean itself and still not have any water to drink. If the Ancient Mariner thought that he had it bad, how about the Modern Mariner who does not just have water, water all around, but it is covering him to a depth of hundreds of feet, and still not a drop to drink.

Fresh water is just as essential to life for the Aquatican as it is for the land dweller. But access to fresh water may be even more problematic. The land dweller obtains their fresh water through a variety of sources, many of which will not be available to the one who has made their home undersea. There are no fresh water lakes or springs to access, and the drilling of an undersea fresh water well is a feat – as far as I can determine – that has never been attempted before. Yet, it is still true that undersea, water is the most abundant and practically unlimited resource.

Ships, submarines and boats all over the world overcome the problem of water usage each day by a few well known processes. Here are the three most used at sea:

Distillation: This is a process requiring a great deal of energy to turn sea water into steam and then condense it back into fresh water, leaving a concentrated brine solution behind.

Desalination by Filtration: In this process, water is desalinated by a combination of filters and pressure. It does not require as much energy as distillation, but it does require periodic filter replacement and there is a comparatively modest energy cost.

Use of Seawater: On ships at sea and on many yachts, seawater is used for sanitary flushing water. There is absolutely no need to utilize fresh water for this purpose. I am certain that in an undersea colony, this time honored and very wise practice will, and certainly should, continue.

Ships and submarines already have large energy budgets required to move their mass over the planet and part of that energy is spent in obtaining fresh water for their crews. On many ships, fresh water is created as a byproduct of producing suitable water for steam boilers which tend to have a much more insatiable thirst for the product than do the human crews.

As a citizen of an undersea colony, you may not have anywhere near the energy budget of a ship because the colony is not normally designed to move. Therefore, selection of the process for obtaining fresh water will be different than on a ship. Specifically, the selection of a fresh water system will probably be elected expressly because of its energy efficiency. Likewise, there may be more than a single process utilized in order to maximize the overall energy efficiency.

An undersea colony will function like a space colony in many aspects of its operation. And one of the key functions in any space colony will be a process called resource recovery. As will be discussed in detail later, the process of resource recovery assures that nothing is lost in the system.   One example of this is condensate as a byproduct of humidity control.

If I step outside my Florida home and look to the west end of my front porch there is a beautiful, healthy Crepe Myrtle plant. If you look right at its base, in the warm months you would see a white PVC line constantly dripping.   That water is from my air conditioning system as it squeezes Florida's oppressive humidity out of the air of my home.   As a total accident, the Crepe Myrtle and the condensate drain ended up in the same place.   But it beautifully illustrates the principle of resource recovery.   Instead of wasting the condensate water, it is used to continuously water the plant – hence, the resource is recovered and reused during the hottest part of the year when the plant needs it the most.

In an undersea colony, humidity control is an essential part of the life support system.   This water vapor is removed from the air and is stored.   It is by definition 'pure' water than can be reused as a part of the colony's fresh water supply – another example of resource recovery.

It is likely that the colony will utilize the lowest practicable energy consumer for recovery of fresh water from seawater. Desalination by filtration, as many yachts use each day, causes a dependence on shore side supply for filters on a periodic basis, which have purchase and resupply expenses associated with them. And yet, the technology is proven and highly developed so that these systems are efficient and reliable.

But the greatest water supply of all is found in re-education and re-training of the aquanauts who come to live in Aquatica. If I were a schoolmaster and could grade the typical human in the average industrialized nation on water usage, I would flunk them with an F minus. And I am not a hard grader! Let me give you an example.

If I were to ask any one of them what it would cost for a gallon of fresh, pure drinking water, almost everyone could tell me it cost less than one dollar. But almost none of those same people could tell me with any degree of accuracy how much it costs to have that same pure drinking water delivered to many rooms of their homes. Let us go out on their front porch and stand by the water hose spigot. Here we will calculate the cost of one ton of fresh, pure drinking water. Based on my monthly utility bill, at my front door, kitchen sink or one of my several bathrooms, a ton of fresh water costs about $1.60. (This works out to less than seven tenths of one cent per gallon!)

Now let us all sit down at the yellow pages and call around to find the lowest cost for having one ton of the vilest, rotten, disgusting garbage delivered to our front yard and dumped. How much would that cost? In my neighborhood, that will cost me around $200! Since custom delivery of vile garbage is 125 times more expensive than custom delivery of pure drinking water, this establishes an unconscious and incorrect mind-set: *fresh, pure drinking water has almost no value at all.* And that is not so much of an attitude as it is a hard fact. So if the toilet is inefficient, or if there is a drip here, or I need to dump 10,000 gallons on my front yard to make it just a smidgen greener, then what does it matter? Water is nearly free. That also translates into very long showers and other wasted water uses without number.

The very first set of instructions for both undersea and space colonists will invariably be the true value of pure water and how to use it properly. Pure water is equivalent to life. Pure water is not easy to come by. And as far as lifelong habits are concerned, they must be changed. Our view of fresh water will have to be reprogrammed so that it is revered in the mind of the user. No longer will it be viewed as free, but it will be viewed as uncommonly precious, because if it runs out, the permanent undersea colony is finished.

Just as water is an *in situ* resource, there are many others available to the undersea colony. These resources are floating all about and lying just beneath them. They are available not only to freely use, but to resell, and therefore have the combined potential of turning the undersea colony into a great empire.

CASTLES IN THE AIR

"If you have built castles in the air, your work need not be lost;
that is where they should be.
Now put foundations under them."
Henry David Thoreau

Let us imagine that you set out to build a home for your family. Your requirement will be to gather up all you need to protect them from the environment around you, to power your new home and to feed your family from the resources at your fingertips. But, instead of being allowed to drive to the nearest lumber yard and hardware store, you and your family have been set out alone on an invisible platform 10,000 feet up in thin air. Your task is to build your family's home, house them, protect them and feed them with the resources you have available in this environment and you are not allowed to ask for any other resource save all that you can harvest from around you. Your only gift is power and any piece of equipment you ask for. It would be the closest thing I can imagine to building not only a castle *in* the air, but a castle *out of* thin air. Now, having been given that assignment, what would your response be? How do you rate your chances of success?

Any sane individual would have to say, even with unlimited power and every imaginable piece of equipment, the chances of building anything under those circumstances are absolutely zero. It is a near certainty that everybody is going to die of not only exposure to the elements but starvation. It is a nearly impossible task.

But what if you were given the identical task on a platform suspended 120 feet beneath the surface of the ocean with the identical conditions – unlimited power and any piece of equipment that you need? Could you do it? Absolutely! All your needs – from shelter to food to more power – are available at your fingertips. In a world where castles in the air are impossible, castles under the sea are most certainly not!

In the oceans, all the resources you will ever need are literally flowing all about you, around you and past you every second of every day. Your task as an Aquatican is simply to know how to capture them. The Third Principle of Undersea Colonization states: "Every colony system will endeavor to utilize *in situ* (in place or on site) local resources to meet every need, from food to gas exchange to energy, in order to reduce dependence on dry land resources."

The purpose of the Third Principle is not political sovereignty, but something far more practical: survival. If the colony depends on a certain resource from the land and that resource is withheld, dries up or becomes too expensive, then the colony's existence is threatened. If a land based community faces the same situation, they have the luxury of waiting it out. Such is not true for an undersea colony. If an essential resource is cut off, the residents of the colony will be forced to evacuate to the surface. They cannot wait for any essential resource.

The undersea environment and surface interface is rich with energy, food, oxygen, fresh water, construction materials, raw chemicals and even metals – all drifting by in the stream of water that surrounds you! The challenge is, of course, how to extract them. Dissolved in every cubic mile of ocean there are about 200,000,000 tons of chemical compounds including elements like gold, silver, magnesium, aluminum, radium, barium, bromine, iodine, sulfur, and many others. It is an environment quite unlike the lands where the surface of the earth has to be shredded to find these elements. In the ocean, all you need is a pump and a process. For every 200,000,000 tons recovered, the environment and its permanent appearance does not change.

Obviously, the undersea resident is faced with the problem of getting to all these resources. Nearly every single process requires specialized technology and energy! While it would take up an entire book to discuss mining dissolved metals from seawater, let us direct our attention to six immediate needs. Four of these elements (not including food and building materials) have been discussed previously in this book, so we will only summarize them as examples of an *in situ* undersea resource:

Energy: Energy can be harvested by current generators and surface swell generators. On the surface, wind and solar energy are also available.

Oxygen: Oxygen is available by compressing surface air or by utilizing semi-permeable membranes that pass dissolved ocean oxygen into the colony. (This technology is in its infancy and is not developed to the point that adequate supplies can be obtained as of this writing.) Oxygen can also be acquired through algae and plant systems inside the colony which would be considered *in situ* supply. There is also the process of hydrolysis which uses electricity to extract oxygen directly from seawater.

Carbon Dioxide Absorption: CO_2 can be absorbed directly into seawater to some significant extent. Again, semi-permeable membranes can also be used to a very limited degree, as can onboard plant and algae systems.

Fresh Water: As I discussed in the previous chapter, fresh water can be extracted from seawater by various methods as well as *in situ* resource recovery methods.

Food: Aquaculture is a very well developed industry and science on the land. But 'open' aquaculture systems in the wild ocean are not so well studied. Typically, nearly all of these applications are limited to open ocean fishing industries. And yet, there is some disagreement as to whether an open ocean fishing venture can be considered 'aquaculture' at all.

There is a wealth of evidence that certain species of fish will become attracted to undersea structures, even in the open ocean. This is true of oil rigs constructed many miles off shore. When any oil rig is established, it becomes almost immediately the home of hundreds of thousands of fish who swim around its structure and find their individual niches. The same is true for sunken ships, some of which are intentionally sunk for the very purpose of creating an artificial reef.

As I can tell you from first-hand experience, the undersea colony structure itself will immediately become such an artificial reef the moment it touches down. Further, aside from the structure of the colony itself, artificial reef formations may be installed aside the colony for the specific purpose of attracting even more fish. Developing a new kind of aquaculture – or fish farming – will become essential to the colonies. Depending on the accepted local national laws or international fishery treaties, these fish may be harvested by the colony for food. By carefully attending to the details of the laws governing the taking of certain species and their sizes, and through proper farming procedures, a perpetual yield may be established by raising the fish in local undersea farms for the purpose of food production, just like any aquaculture farm ashore.

I fully realize that this discussion may lead to some sensitivity. And yet, the undersea colony is quite unlike a fishing ship – it cannot move and it cannot by definition deplete its own fish population or the food resource will stop! Therefore, the undersea colony is bound by its own survival to carefully manage its aquatic populations to its own peril.

Such a project is rich with scientific potential. By establishing large artificial reefs and by careful management linked to their own survival, the colony is obligated to keep one of the most exhaustive and vigilantly researched studies of ocean fish species ever conducted in the undersea environment. Rather than try and investigate these species from the surface, Aquatican marine biologists will be in place and able to monitor each day with no end to their research window – something never before accomplished.

Hence, undersea colonies will provide a service to the entire planet by gaining in depth knowledge that cannot be duplicated from the surface. They will gather information that benefits our entire planet – especially the environment and the endangered aquatic species that have been over fished and hunted down by the countless millions by floating fish factories.

In Aquatica, only fish that are approved by the scientific and fishery officials will be taken. By definition, in this process, not a single species will be taken by 'mistake', then killed and tossed aside as a 'waste fish' as is so common on the surface. In Aquatica, preserving the environment and its balance is equivalent to life and permanence. Because, if it is altered by our presence, then the environment itself will force us to leave.

Building Undersea Structures With Seawater and Sunlight

In 1981, Wolf Hilbertz patented a remarkable new technology whereby he demonstrated that he could build solid structures out of raw seawater combined with a low voltage electrical current. The invention's formal title is: "Mineral Accretion of Large Surface Structures, Building Components and Elements," U.S. Patent No. 4,246,075, (Jan. 20, 1981). Hilbertz followed this patent up with two more in 1984 and 1992. After the revelation that such a feat was in fact not only possible but relatively simple, four more patents followed using what had become known as 'Biorock' and 'Seament' production.

In 1990, artist Chris Scala patented the process whereby he used the accretion technology to produce in-water sculptures. Chris Olstad and Scala teamed up at the Marine Resources Development Foundation and began to create sculptures out of seawater using low voltage current. In this process, a shape is built out of wire mesh, dropped into the ocean, and a relatively low power DC current is run through the wire (Chris uses solar panels to power his creations). On one side is the cathode (- post) and the other is the anode (+ post). As the current runs through the shape, electrolysis is established – oxygen is generated on the anode and hydrogen at the cathode in the expected two to one ratio.

Obviously, the oxygen produced as a by-product from this process can be captured for breathing by the human population and the hydrogen for fuel cells and energy production. But a very fascinating thing is happening at the same time on the wire grid! Solid materials are being deposited on the wire mesh made up mostly of an ever thickening layer of calcium carbonate. Soon, the spaces between the wires close in and fill up, making a solid structure.

Accreted structures made from this 'Biorock' or 'Seament' in time become solid enough to hold air, as Chris Scala discovered. They can become so solid, in fact, that one sculpture even held a large pocket of hydrogen gas – perhaps the most penetrating gas known. Hydrogen gas, H , the lightest of all gases, consists of the smallest molecules. This is what gives it unprecedented penetration ability, even through materials as dense as concrete. Yet Scala's accreted creation was dense enough to trap even hydrogen gas!

Thus, out of power generated by solar panels or any other source, undersea structures can be constructed in any shape imaginable, machined much like concrete structures are on the land, and then be filled with air. These structures can be created out of the seawater flowing by – truly a kind of 'castles out of the air'! Further, while the structures are being formed, the colonists can harvest hydrogen fuel and oxygen to breathe. And all of this natural construction can occur at the undersea colony site, powered by the sun, wind or waves!

Shown in the photos on the opposite page are a series of close up shots of a completed Seament structure accreted undersea and energized wholly by solar power. These items have not been machined and this is exactly what the natural process looks like after a set time of accretion has been completed.

SHOWN HERE IS THE ACCRETED SURFACE FROM ABOUT 6" (L)
AND 2" (R). NOTE THE ROUGH AND UNEVEN SURFACE TEXTURE.
ALSO NOTE THE UNDERLYING "FABRIC" SHOWING THROUGH THE
SURFACE LAYER IN THE RIGHT PHOTO, BOTTOM CENTER.

SHOWN HERE IS AN EXAMPLE OF HOW THE FABRIC CAN BE
SHAPED IN ADVANCE AROUND ANY STRUCTURAL COMPONENT
ALLOWING THE SEAMENT SKIN TO LITERALLY GROW AROUND IT.

The accretion process does not occur overnight and the structure shown above required more than a year to produce. One study suggests that one kilowatt hour of electric power will result in the accretion of 4.2 pounds of Seament (4.2kg/kw-hr), but that data has not been repeatedly verified. Another study found that only 0.046 kg/kw-hr could be produced. There is precious little data available, and there are still many experiments that would be desirable before the technique is fully refined. Most certainly, the first Aquatican colonists will 'build' on these ideas immediately. If the power is relatively 'free' and all they need is time, the initial accreted structures will only depend on setting up the 'fabric' framework and turning on the power. In a year or so, the colonists will have solid structures to move in to!

In all of this discussion about *in situ* resources, the most important consideration is how the activity affects the ocean environment. The undersea colonist can literally create his empire out of materials dissolved in the sea or by the exclusive use of renewable resources. But in the end, every decision is based on the effect of the human presence on the environment and by asking the question, how does the human presence make the one planetary ocean called Aquatica a better place specifically because of man's presence?

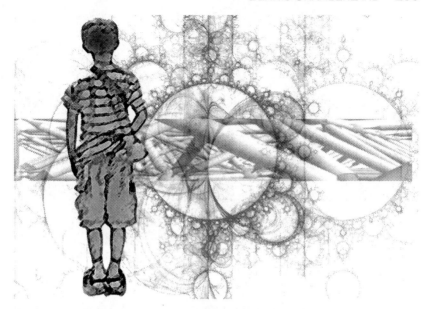

ENVIRONMENTAL STEWARDSHIP

"The sense of all these things comes to him most clearly in the course of a long ocean voyage, when he watches day after day the receding rim of the horizon, ridged and furrowed by waves; when at night he becomes aware of the earth's rotation as the stars pass overhead; or when, alone in this world of water and sky, he feels the loneliness of his earth in space. And then, as never on land, he knows the truth that his world is a water world, a planet dominated by its covering mantle of ocean, in which the continents are but transient intrusions of land above the surface of the all-encircling sea."

Rachel Carson

T he Fourth Principle of Undersea Colonization: "Every colony system will set out to reduce their impact on the environment to zero, attempting, in as far as possible, to become a net positive environmental impact specifically because of its presence."

This is the most important chapter in this book for two reasons:

<u>One</u> – if humanity is to occupy the undersea regions of the world, from the advent of the first human occupant, we must ensure that we will not only be responsible environmental stewards, but that we will make certain that we enter there with a whole new environmental conscientiousness that is based on an advanced philosophy of environmental understanding and responsibility.

<u>Two</u> – we must soon awaken and recognize that if we do not enter Aquatica as permanent resident stewards, that we as a species have failed in our responsibility and we simply have no right to be there as we have no legitimate business there. We have a clear and unambiguous duty to ensure our presence in one of the world's most important ecosystems: to closely monitor it and cease leaving the hit and miss observations to machines and woefully limited robotic systems. Our prime mission objective must be for us to be there to make things better, or we should not be there at all.

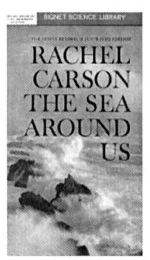

As a young college student, I purchased a book in the campus bookstore titled, *The Sea Around Us*, its cover shown here. I clearly remember the day I purchased it, where it lay on the shelf and carrying it back to my dorm room. I remember all those things because of the impact it was to have on me. At the time, I had already been dreaming about an undersea colony, so the book was the perfect companion to that dream. What I could not have known as I paid for it was that it would become more than just a companion.

While Rachel Carson had long since passed on, she waited patiently between the covers of the book to teach me an essential

RACHEL CARSON

idea – that the sea is not simply an empty void to explore, but it is a living creature that forms the very blood of the earth. For less than $2.00 the thoughts and ideas of a gentle but brilliant woman were to be my faithful companions and teacher over these many years.

While best known as the author of *Silent Spring*, Rachel Carson penned two books prior to her magnum opus: *The Sea Around Us* (1952) followed by *The Edge of the Sea* (1955). What these books have meant to me underlies a good portion of our LNW team's motivation to enter the sea. We are to be caretakers of this magnificent and enormous void. We are entering not to alter but, instead, with the prime objective of watching, monitoring and focusing the intelligent eye of a dedicated cadre of scientists, ecologists and teams whose primary objective is to understand and to protect.

It is a wonderful experience to sit quietly and read Rachel Carson's works. Indeed, as a young man, I clearly remember sitting on the rocky shores of La Jolla, California, on many foggy mornings reading and rereading my well worn copy of *The Sea Around Us*. I will never forget the interconnectedness I felt as I read – as I heard her voice speaking to me though the pages. She said of this very moment, as though voicing some precognition: "If a child is to keep alive his inborn sense of wonder without any such gift from the fairies, he needs the companionship of at least one adult who can share it, rediscovering with him the joy, excitement and mystery of the world we live in." In so many ways, I was her child at the beach, and she the learned adult teaching me on that timeless and beautiful shore.

As I read her thoughts, I remember the sound of the waves and the gulls and the smell of the ocean. I learned that the task ahead of me was not so much engineering as it was one of guardianship. The engineering was to make possible *the way* – the guardianship was to become *the why*. As I read, I realized that this little book and her others were teaching me a philosophy that would drive my engineering of the undersea structures of man. Like the living reef and the symbiotic relationship of all the sea's creatures, I would build my structures to match and complement nature – not to dominate it; not to ruin it; and certainly not to compete with what God has already engineered as original perfection.

Carson wrote: "To stand at the edge of the sea, to sense the ebb and flow of the tides, to feel the breath of a mist moving over a great salt marsh, to watch the flight of shore birds that have swept up and down the surf lines of the continents for untold thousands of years, to see the running of the old eels and the young shad to the sea, is to have knowledge of things that are as nearly eternal as any earthly life can be." She taught me by symbiosis through those pages, like the invisible but comforting mentor that she was, "There is no drop of water in the ocean, not even in the deepest parts of the abyss, that does not know and respond to the mysterious forces that create the tide."

All of those non-living forces and all of those living creatures were irrevocably bonded together one to another – and all moved in one way or another – toward life or away from life.

And then she said of my time with her through those almost magical pages, as though smiling to me from the grave, "The more clearly we can focus our attention on the wonders and realities of the universe about us, the less taste we shall have for destruction."

But, to our shame, humanity has trashed and raped our environment to the point that it is becoming more and more difficult to peacefully coexist with the natural order. In so doing, we have polarized our own culture to the point that it is nearly impossible to have a coherent, unemotional discussion of the impact we are having on our own planet. The two 'sides' in discussion are so antagonistic that it requires a miracle to be able to sit down and discuss these ideas rationally and without disgust for one another.

For a brief moment, let us together assume that miracle has occurred. No matter what your position on environmental issues, may we sit together in peace for just a few pages and discuss the environment in reference to undersea colonies? There are two very polarized sides in this debate, and to keep from insulting either by an unintentional misrepresentation, let us just agree that there are two sides and go on without making the mistake of attempting to define them or to reconcile. It really does not matter anyway, because any reasonably intelligent person already knows very well what they are.

I have a dream that the very process of undersea colonization can be used as a teaching tool for everyone on the planet. It is my dream that the establishment of these human cities under the sea can teach one very important thing: that humanity can have a positive rather than a negative impact on the environment. Rather than going there and depleting, mining and trashing it, we can go there and actually improve it. And, because we are there, the planet will be better off than if we had never arrived.

Walt Disney had such a dream and his turned into the model community environment which he called the Experimental

Prototype Community of Tomorrow (EPCOT). Such a dream is also possible as an undersea colony. Walt never saw the full extent of his dream come true because his original idea was altered into a community that allows visitors to tour every day instead of a community of families in a real, functioning, idyllic city. In many ways I am very happy at the way things turned out because EPCOT is one of my favorite places to go and I spend many hours on quite a few days of the year experiencing one part of his fantastic vision. As fate would have it, EPCOT even includes a model undersea base! A city like EPCOT that is fully shared is a wonderful idea.

In my dream, each undersea colony will start out on its first day having a net zero impact on the environment, but soon in the process the impact will become positive. Further, the community will be a zero waste society and will teach that all we now consider waste is actually a community resource. That concept is developed in a chapter just ahead.

Because of humanity's poor record of environmental stewardship, this concept is almost unknown. And yet, it is actually very possible! In my years of work with advanced life support systems being considered for off-planet applications such as Moon and Mars bases, I was tasked with understanding the big picture of the bases and colonies on other planets in order to comprehend how my piece fit in. In so doing, I was a part of a larger team that pieced together conceptual communities that had a zero impact on their surroundings. In space there is nowhere to go to get other resources, so even 'waste' becomes a valuable resource!

In the oceans, these principles apply just as fully as in space or on another planet. And there is no better way to model this to our planet than in an undersea colony. The ultimate 'spin-off' of any technology is a powerful, life changing or planet changing philosophy and not just another gadget. And this philosophy is indeed powerful enough to change the course of all of our futures, across this planet as well as on others.

This new philosophy teaches five fundamental principles:

1.   We have a right to be on our planet and to live here, to reproduce here, to enjoy life and to pursue happiness and enjoy personal freedom.

2.   We have a responsibility to preserve and protect our environment that shares an equal weight with our right to use that environment.

3.   Each individual shares equal responsibility to leave our planet better than when he came.

4.   We have a clear accountability to clean up after ourselves and not bury our wastes for another generation to clean up after us, because –

5.   *There is no waste.* By definition, all wastes are resources that must be recovered and reused.

Granted, there is a lot to learn and unlearn here. For example, if we look closely at principle number five, new technologies must be developed to go along with that specific philosophy before it can totally work. But these principles are an important first step toward changing our understanding, life philosophies and behaviors before it is forever too late. One of the primary purposes of the undersea colony is to teach by demonstration this philosophy and to be able to point to a community of humans who live these principles in an organized, functioning community – people who actually occupy a new territory and make it better expressly because they are there.

It is true that one of the reasons we are at such odds with one another over environmental stewardship is that many of us have different expectations about what that term even means. At the far end of one side of this ongoing cultural war, some speculate whether humans have a right to even live on the planet at all, much less expand their presence! At the far end of the other side is the belief that everywhere a human foot trods is theirs to use or abuse as they please for any purpose they want.   But there is always extremism on any side of a discussion that will never accept new ideas or approaches that actually have the capacity to work.

Yet, this book specifically speaks to a part of that extremism! In a time of deep concern, there is an opportunity for a new idea to emerge that will solve a planetary problem through a fresh look at some possibilities that have never been considered before.   Extremism in human communities has been used to humankind's advantage before, because it represents a kind of high energy leverage in a philosophic debate.   And that high energy is typically required to kick off new ideas and approaches that normally would have lain fallow.

Just as important as entering the ocean as permanent residents for the first time is our implicit *responsibility* to do so. There is no excuse for our timidity or our intellectual failure to recognize just how crucial it is for us to approach this task with all haste.   Jacques Cousteau pointed out to us in his extraordinary undersea ventures how the ocean environment began to degrade just in the brief time he had examined it.   Since his well documented journeys, no other comparable eyes have taken his place with such a consistent view of Aquatica.   We are now effectively blind to what is occurring beneath its mirrored surface.

While it is true that there are a few probes here and there taking readings for widely varied reasons, there is no consistent presence in the sea that, unlike remote probes, views the environment with intelligent eyes and highly sophisticated brains as a coherent system.   Whole communities of aquanaut Aquaticans will seek answers to questions so essential that the well being and environmental health of the entire planet will depend upon their results.

When the first undersea colonies are established, they will immediately begin to make consistent observations of the health and well being of the undersea regions of the earth and its inhabitants as a priority activity.   Within a few years after this pursuit begins, a baseline of data will be recorded against which the health of the oceans may be determined consistently and effectively from that point forward.   As of this date, there is no baseline because there is no permanent presence to record the data day after day, year after year.

Ironically, we have abandoned our clear and unalienable responsibility in some cases because we claim that just by going there would in some way make matters worse. That argument is akin to saying that just by leaving our homes each day to make any environmental investigation that we have somehow increased the harm of the problem we set out to investigate merely by the act of examining the problem. That argument is ludicrous, of course, but it is a result of a lack of education and awareness that has turned a credible scientific discussion into one of fear – a fear of invisible evil and superstition reminiscent of pre-renaissance eras, all without logical justification.

The planet's environmental woes – above and below the seas – will never be solved by invoking nameless fears and attempting to join the questions of legitimate science with political correctness. Philosophically, science and politics mix like oil and water. And if they are mixed, the result is anything but scientific and typically resembles a Frankenstein-like monster of hideous, out of proportion pseudo-logic that benefits no one, least of all the environment of the earth and all her creatures large and small.

What We Are Looking For

Our League of the New Worlds team is planning to establish a permanent undersea colony off the central Florida coast. The large manned habitat that will become the hub of the permanent colony there has been named *Challenger Station*, christened after the 19th century British oceanographic ship, the *HMS Challenger*, which made so many great ocean discoveries. Its principal duty will be to collect the first consistent, permanently ongoing baseline of ocean environmental data ever accumulated in history.

The colony that will ultimately surround the *Challenger Station* is planned to be an undersea colony of science teams that will investigate many environmental parameters as the major part of its duties there. Above the colony will pass the mighty Gulf Stream that will allow continuous sampling of the largest ocean current system in the world.   We will be able to analyze continuously some 35,000 miles of ocean and everything that this great undersea current carries in its flow each year as it sweeps through the North Atlantic Ocean and across the whole of the northern hemisphere and back.

Here are some of the things we will be investigating as we work to build the environmental data baseline:

Chemical analysis: Everything released on land ultimately makes it way to the great planetary ocean. Said Rachel Carson, "For all at last returns to the sea – to *Oceanus*, the ocean river, like he ever flowing stream of time, the beginning and the end." The ocean is a great mixing basin for chemical species that are poured into the seas from the nations that adjoin the great currents. The Gulf Stream will yield a laundry list of man made chemicals from both Europe and the United States including industrial chemicals as well as pesticides and artificial fertilizers. While it is more than just tragic that this register has not been built before, we will at least begin to understand and compare this directory as we view it from year to year.   These chemicals, as one might suspect, mix in with the natural environment and can have diverse and potentially devastating effects on every aspect of marine ecology, including reef and fish populations.

Microorganism analysis: The ocean's first level of the food chain is the population of invisible aquatic microorganisms.   Their population dynamics affect every level above them. A study of this dynamic and highly robust population is essential to understanding every echelon above them.

Plankton analysis: Plankton comprises the greatest single food mass in the ocean and remains the greatest single generator of oxygen on our planet. Cataloguing these species and understating their balance is essential to a comprehensive understanding of the ocean environment.

Wildlife analysis: The wildlife in the Gulf Stream includes many diverse species from whale to dolphin and scores of fish and shark.

This population dynamic is also vitally important to catalogue and understand in a continuous study.

It is true that on land we have built a massive and complex culture.   But, unfortunately we have allowed that culture to dramatically affect the undersea regions of our planet.   We have not even the slightest clue as to what damage we may have already done or how to reverse the negative effects.   However, there is a team preparing now to go do something about that problem – and we need your support!

One of the most important things that we have to teach the culture of the land areas is a philosophy of resources.   Instead of leaving behind a legacy of waste, we intend to leave behind a legacy of preservation through a process of environmental management called resource recovery.

RESOURCE RECOVERY

*"The men who are not interested in philosophy need it most
urgently; they are most helplessly in its power."*
Ayn Rand

I t was my first day on the job as an advanced life
support systems engineer. My boss called me into his
office and gave me my first assignment. He asked me
to come up with several suggestions for the processing of waste in
a space colony that utilized a bioregenerative life support system.

I went back to my office, poured over the details of the
system and gave it several days of intense review and focused
thought. I thereupon returned to my boss and told him, "I'm
sorry, sir, but you don't have any requirement for a waste
processing system."

"Why not?" he responded, taken completely by surprise.

"Because you don't have any waste."

He stared back at me for a moment, and then smiled as
though I were joking.

But I was not. I explained to him that in my view, an advanced life support system of any characteristic, by definition, cannot generate a waste stream. Every material created by the system or the humans, no matter how apparently vile, was not a waste at all but an essential system resource. He wholeheartedly agreed and instead of focusing on a 'waste processing component', attention was directed to what was thereafter known as the 'resource recovery system.'

I have stressed throughout this book the power of philosophy. Here is another case where the philosophy of the system drives the system design and fixes human expectations. In this case, the old philosophy of segregating and then categorizing any material as a 'waste' invariably led to a process of 'throwing it away'. Along with those expectations came others to match – the invariable attendant reasoning that the generation of waste was an expected norm not only of the culture but of the individuals within the culture.

Unfortunately, throwing a waste away in any environment is not actually possible. A culture may drive it off to a distant point and cover it with dirt or even more disgustingly, dump it on the bottom of the ocean as is frequently done – but they have certainly not actually 'thrown it away'. They have only moved it. It can never really be thrown away although they have conveniently convinced themselves that they are doing just that. It is a dangerous false assumption that immediately threatens the lives of countless species, including our own.

The problem with false assumptions is that the truth sooner or later becomes evident, and usually in the form of a shocking revelation. One day someone lets the city commissioners know that the waste they had discarded and forgotten is now in their drinking water in a whole new, insidious form. And the irrigation system from the local streams and wells is spreading it all over their food crops. Ultimately the fish crops all over the planet are likewise contaminated.

The community will be doubly horrified because they had always thought of themselves as environmentally astute. They had in place the most robust and effectual anti-littering program on the planet. They had very rigid building codes such that their city on the outside looked magnificently clean and beautiful. Unfortunately all they had done was engage in a pointless exercise in external visual appeal. But inside they were rotting because they had tried to hide their waste instead of deal with it. In fact and in reality, their own waste came back to haunt them because they could not hide it forever. Any waste can never stay hidden for very long.

In an advanced life support system, all waste must be dealt with elementally and chemically because it is the resource of tomorrow. Even in human waste is locked the carbon required for tomorrow's crops. In an advanced space system, if you throw away human waste or foolishly try and hide it, the system will rapidly wind down, run out of carbon and everyone will eventually die. This powerful philosophic tool is essential for our own space system – the earth – as it spins and rides through the cosmos. Somehow we must relearn our philosophies and retool our culture or we will certainly die in our own refuse.

The undersea colony is a perfect place to teach all our human cultures how we can accomplish these things. Here is an environment that is suitable to require such a philosophy and to implement it immediately with the very first outpost. The planet's oceans will be populated by a culture of Aquaticans whose going-in philosophy is resource recovery and the cultivation of a human awareness among their people that the environment is one of eternal importance, not just to this generation. We will teach to everyone who will listen that we invest in the environment from generation to generation and not mine it for our own. We will demonstrate how it is possible to actually make it better from one generation to the next instead of building a deficit that soon no one will ever be able to repay.

Resource recovery is a highly complex scientific and engineering process. It is one that we have only now begun to understand. To be most candid, if we wanted to implement a fully operational resource recovery program today, unfortunately it would not be possible. If there were ever a reason for any society to invest an Apollo-type effort crash engineering program to literally save us from a next generation disaster, this would be it. For the amount of money spent each year on Easter candy (about

$1.9 billion USD in 2007), we could literally work miracles for our planet – to the benefit of our children and theirs.

However, there are major pieces of such a system that we are able to implement, and the undersea colony could immediately be designed to incorporate all those elements. As two of many examples, water exhaled by humans and even urine can relatively easily be recycled into fresh water. While we cannot recover all human waste, more than 99 percent of its mass can be reduced to a fine ash and nearly all of its water mass and CO_2 can be recovered. But the human waste stream, while most have far overblown its actual significance in their minds, is a miniscule waste when compared to the wastes that truly wreck havoc on the environment.

Heavy metals and complex organic wastes that are the byproduct of many industrial processes are the true villains. Before the late 1970's these chemicals were literally poured down wells, dumped on the ground or flushed into municipal sewers and even into pristine rivers and streams.

As a boy I spent countless hours playing by my stream at the rear of our property. It was truly a magical place full of crystal clear water that danced and curved along its length over waterfalls and rock pools that held and nourished many diverse creatures. I loved my stream and often sat by its banks for hours nearly every day with my faithful dog Max by my side. My stream held a magical, nearly mystical place in my heart as a boy.

One day I walked down to my stream and, even before I reached its banks, I could smell an odd odor. As I approached the wooden bridge I had built across it, the odor became overpowering. When I looked over the bank and into the water, I could see it was no longer a beautiful, crystal stream, but instead it was a slowly moving river of black, putrid petrochemical waste whose odor was so strong that it brought tears to my eyes. Max sneezed and ran away!

My stream was gone, replaced by this black mass that had killed every creature that could not escape it. I ran home and brought my dad back to the scene. He stared at it with disgust.

"What is it?" I asked, horrified.

"Someone unloaded their truck into the stream," he replied.

"What truck?"

"A waste hauler."

"Why?"

"Because it costs them money to take it to a disposal company," he explained. "They are paid to pick up the waste, which included the disposal fee, but they dump it in a stream and keep the money."

"But they ruined my stream!" I complained. It was the very same stream in which I had built so many underwater insect colonies and had camped by and had fished in for crawdads and frogs.

Dad just looked sad and nodded his head slowly. "Don't worry, son. It'll come back."

It eventually did come back as the rains and snow melt washed it over and over. But it took more than a year and not every creature I was familiar with actually made it back. I am certain that it never fully recovered to its former self.

The deep anger I felt at that waste hauler has lasted a lifetime. His crime was ultimately a crime of deep selfishness, greed and deceit. It was a one man decision, a one man operation that had killed so much and ruined a beautiful part of the earth.

Every time we make very small decisions that take away even a small part of our delicate planet, we stand in the shoes of the waste hauler. He apparently had no compunction against opening the valve on his truck and flushing the nameless stream with his dangerous chemicals. He did not care about its wildlife or its special connection to a boy; he only cared about himself. Each time we toss something away, we make the decision to move it and hide it so that some other generation will be forced to deal with it. Each time we contaminate our planet, we kill a part of life and we destroy a dream.

It was not the waste hauler who produced the waste. Had the company who produced it had a method of dealing with the very waste they generated, or a process of turning this chemical soup into something useful, the waste hauler would have been forced to find a more honest line of work. Had we all lived in a culture of 'resource recovery' instead of a culture of 'waste processing', my living stream and all of its precious creatures would never have been wiped out in a single horrible act.

That is the view and the vision that Aquaticans will go to the oceans of the world embracing. We go there to heal, to clean, to recover what we use, to change the way we as humans have learned; not just to do better but also to do things differently. As Rachel Carson so aptly stated, "Like the resource it seeks to protect, wildlife conservation must be dynamic, changing as conditions change, seeking always to become more effective." We will change our view of the earth and ultimately leave our planet improved, not only under the sea but also on the land.

As we approach that time, there will be many obstacles. It is most surprising and ironic that some of those who seek to keep us from going below do so because of their zeal to 'protect' the environment from us! As we prepare a powerful and perfect classroom and platform for teaching a commanding new philosophy to our fellow humans as the fundamental reason for why we do what we do, many of our major detractors are those who are unwittingly trying to prevent us from teaching this new philosophy of preservation. Paradoxically, in the end, sometimes those who say they carry the mantle of protection actually end up in the shoes of the waste hauler.

As we look ahead to the time when men are finally more concerned about the earth than themselves, populating Aquatica will require not only environmental stewardship, but sound financial stewardship and commitment as well.

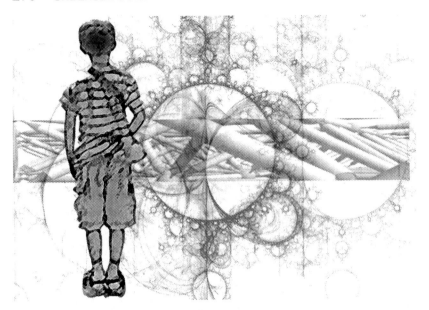

ECONOMICS

"No bucks – no Buck Rogers,"
The Mercury Seven Astronauts

T his statement was reportedly actually said by the original United States astronauts to the engineers building their spacecraft. It is a five word sentiment that carries with it great power and ultimate truth. Every venture must have an income stream or the venture remains locked up as a mere figment of someone's imagination.

If the Mercury astronauts wanted a window in their spaceship and there was no funding to install a window, then they would have been forced to make one of three decisions: ride in a spacecraft without a window; quit being an astronaut and go do something else; or find the money for the window. The program decided the latter idea was better than the first two. In the case of the United States Government, the decision is not as problematic. If the 'company' runs out of money, they simply borrow from themselves or print up some more. No one else has it that easy.

Every expedition or new venture has to find 'start up capital' or 'venture capital'. Most receive support from investors that have some hope of seeing a return for their outlay. There is high risk capital that is always associated with the possibility of a high rate of return, and so forth.

I particularly enjoy the economic story behind the voyages of Christopher Columbus. He was looking for funding for what he called "The Enterprise of the Indies". He would require three ships and crews to find a new route to the Indies that he was convinced was there. After failing to convince King John II of Portugal to back the plan, Columbus followed the court of King Ferdinand and Queen Isabella of Spain for ten years seeking a gracious audience. Finally, after befriending a court official, the bureaucrat whispered to the Queen that Portugal was embarrassing Spain in their many explorations and, in the end, Columbus's request would cost less than a single royal party. The queen finally agreed and the result is well known.

Every new exploration venture requires an economic venture capital or it never gets off the ground. The first and subsequent undersea colonies will also require such start up funding. Exploration ventures are sometimes capable of receiving government assistance in the form of designation as non-profit entities. This permits corporations and individuals to contribute money or products to the venture for a tax incentive. Often the expeditions agree to post the corporation sponsor logos on their vehicles or uniforms as a form of advertisement for the donor companies. A tax write-off and subsequent advertisement can prove to be a powerful incentive.

Product contributions are quite essential and even irreplaceable because they substitute for cash for many expedition needs and greatly reduce the requirement for funding. But regardless of the many sources of assistance, every expedition ultimately requires cash. It is simply impossible to manage a large expedition, such as the one that will be required to establish an undersea colony, without a source of funding.

In one of our expeditions, many of our major costs were unknown to us the day we left home. There were immediate and daily requirements requiring cash outlay for things as small as nuts and bolts to things as large as unexpected equipment breakdowns that required immediate replacement. Were it not for the availability of cash to pay for these items, the expedition would have stalled in the field; a terribly unfortunate fate that happily was not ours.

In the earliest stages of any expedition, including the quest to establish permanent undersea colonies, there must be the development of a broad, multi-phased economic strategy. The first phase is building the funding documents that will provide the fundamental communications package to potential supporters. The next phase is assembling a list of potential product donors, followed by assembling another list of potential cash providers. Once these milestones are accomplished, the requests for support are sent out.

This is a very harrowing and complicated phase. The expedition's leader knows that if there is inadequate support, the project can never proceed. I have a friend who worked this phase in maddening cycles for 16 years before he finally found enough support to begin his expedition. During those 16 years, he found an amazing number of supporters but there was never enough accumulated support to push him over the edge and make it happen. Ultimately, by incredible self sacrifice and cutting enough corners, he was finally able to put it all together and launch his expedition.

This process underlies the hazard in establishing project launch dates in advance. While launch dates are important for scheduling seasonally sensitive projects, they come at a risk. If the expedition fails to make its date due to lack of sufficient support, it appears to be a potential weakness or even a failure. On the other hand, a hard date is a wonderful and powerful psychological tool that gives everyone involved a sense of confidence and a hard target to aim for. It is so much more effective than "we plan to do this... sometime in the future..." which becomes a real disadvantage when attempting to attract potential sponsors.

The support of any expedition must include products, cash and volunteer man hours to proceed to the next phase, which is assembly of equipment and new technology. Regardless of the number of volunteers, this can only be accomplished if both product contributions and cash are in place even as the volunteers are working hard.

Please be aware that all of these phases in every expedition are driven strictly by economics. While ideas are the engine that creates the expedition, economics is the fuel in the tank. Regardless of how powerful your engine is designed to be, without the fuel it is pretty much useless.

At this phase, as the support begins to appear, the expedition may be given a green light for building and assembling equipment and training personnel if the expedition's leader is confident that things will continue at a level that is necessary to attain Mission Day One. But until the final support is received, the expedition is still not assured.

As we view this process by which most 21st century expeditions are compiled, we recognize that it is primarily one of assembling venture capital. But the 'investors' invariably choose to support for altruistic motives, since their return is principally a tax incentive. The significant distinction in this discussion is that most other expeditions are one time events that have a start and end date. Our venture is far different. We seek to establish a permanent human presence in the oceans. There is a vast dissimilarity between an expedition that may last for weeks or months and one that has no scheduled endpoint. In terms of the expedition leader, he is then forced to add another, far more uncommon and complex component to his economic planning: permanence.

It is here that few expedition leaders alive in the 21st century have dared to tread. Climbing a mountain, sailing around the world or hiking across the poles is one thing. But in the end, the explorer gets to come home, close the books and go back to his profession. Not so in our venture. Once it begins, it must have a steady flow of economic support or it will fail to meet its stated objective or permanence.

In the case of the wide-ranging classes of expeditions that founded the new worlds in the 16th century, there was a broad spectrum of economic potentials, each underwritten by high risk venture capital. It is very important and interesting to point out that the European governments were not the exclusive drivers.

Private companies such as the Virginia Company, London Company, Plymouth Company, the Society of Merchant Venturers and others were heavily involved in financing these expeditions for their own economic interests and gain.

Yet, eventually everything leads to yet another phase in the venture: economic self sufficiency.

As we look forward to establishing the first viable undersea colonies, we must look beyond the first-phase altruistic and philanthropic start-up sources to the self sufficient economic engines that will drive the economic foundations of the new colonies. At some point beyond the establishment of the undersea colony, an economic transition must be made: the community will be forced to fund itself – to sink or swim, if I may be allowed the pun.

There are many sources of possible income for the new colonies and most of the prospective economic potentials are no more known to us today than they were to the colonizers of the new worlds half a millennium ago. Obviously, the undersea colonies will establish laboratories in the very first modules launched. This will permit not only local Aquaticans to make their daily investigations, but availability will be shared with visiting corporate and academic scientists who will be levied a daily per diem to study there.

The colonies can be used as a source of undersea data generation of every conceivable data type, from physical oceanography to marine biology and environmental status, which is unparalleled in history. While much of this data will be freely published by scientists in refereed journals, the colony will decide which data has a financial value that will enhance their economic viability and ultimately their survivability.

Governments may find many uses for a permanent undersea colony from research applications to space analog studies. It is also possible that corporations will continue to sponsor undersea colonies to keep their logos prominently positioned on the uniforms and colony structures as a viable continuing advertising incentive from year to year. This approach has worked exceptionally well with sporting figures and NASCAR teams.

Like the International Space Station, undersea colonies will eventually allow visitor-tourists to fill a certain number of pre-designated cabins to experience the life of an Aquatican for a fee. Undersea visit packages may be advertised in many different venues such as:

The Tourist: Here the visitor is brought down in a submarine and given a tour of the facility and permitted to stay for any number of days.

The Undersea Adventurer: In this package, the visitor swims down (where possible) to the colony from the surface with a guide. He is allowed full pre-planned access of the facilities in a number of tours outside as well as guided undersea scooter rides around the complex and nearby artificial reefs. There will also be ample opportunities for photography and observation of local sea life. He may even participate in some colony support functions such as food harvesting, equipment transport or structure maintenance.

The Private Scientist: The visitor is allowed to bring their own pre-designed experiment or to watch local scientists as they conduct their work. If desired, the visitor may be permitted outside to view or participate in experiments in progress.

The Ecologist: This visitor is focused on the colony's environmental work and interfaces. They may or may not wish to venture outside for a look at the artificial reefs and other environmental monitoring stations.

The Reflective: This visitor wishes to be transported to the station and then left alone to write, meditate or just look out the windows.

The Romantic: Here, the visitors are allowed to spend time with one another as well as take part in any tours that are available. It is perfect for the ultimate get-away honeymoon. Marriages can also be preformed on site by the colony's Mission Commander or Chaplains.

<u>Create Your Own Adventure</u>: Here the tourist can combine many different parts of the other packages to create their own adventure.

<u>Check-box Options</u>: Each visitor on any package is allowed to pre-determine certain aspects of their visit from options of how they arrive and depart (submarine or swim) to size of stateroom, to dinner at the Mission Commander's table, etc.

Considering today's entertainment oriented society, it is probable that an undersea theme park and all of its associated aspects will eventually be partnered with a colony. The way will also be paved for the movie industry to take advantage of spectacular locations for filming and new venues for television shows and reality series will be created. In light of these broadly based economic possibilities and potentials, venture capitalists may wish to invest with a contracted return up front, even before the first colonies are launched in the same way as did the first companies who invested in the new worlds.

The economic picture of the undersea colony is as bright as in any other historic venture. There is a clear and unambiguous economic plan that has been effectively employed in past venues and will certainly work again when approached enthusiastically, intelligently and in a spirit of radical creativity.

Yet as the colony grows and the economic picture evolves into certain self sufficiency, the colony is still in the wild ocean and must always be on the lookout for the various real-world scenarios that may ultimately pose a threat.

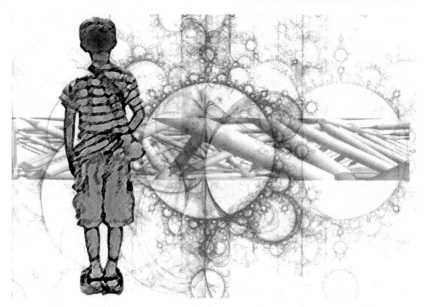

THREATS

*"Nothing at all will ever be attempted if all possible
objections must first be overcome."*
Samuel Johnson

I have heard many objections to establishing an undersea colony, and nearly all of them are based on some level of fear. I even heard an objection from a space advocate who worried that if an undersea colony flooded that "everyone would drown like rats." And yet that individual had no expressed fears about the breach of a paper thin space colony wall by a micrometeorite or piece of space junk traveling at a relative velocity of Mach 36. Radical decompression in vacuum is just about as undesirable a death as drowning and any sized particle traveling at supersonic speeds – even a fleck of paint – could leave a good sized hole in the human or his spaceship.

The point is, there are many ways to die either on the surface of our planet, in space or undersea. The idea is never eliminating risks, because that is impossible. It is not even 'minimizing risks', because if that were our process, none of us would ever get into an automobile or into an aircraft. The idea is to understand risk and then *apply* a learned reduction of risk potential to our activities.

When I get into a car, I fasten my seatbelt. I do not drive impaired. I pay attention while I drive. I then watch out for people who are impaired or those who are talking on cell phones and not paying attention or other risky driving practices. When I fly I pay attention to the safety instructions and look for the exits before the plane sets up for takeoff. When I SCUBA dive, I hold a safety briefing with my dive partner and will, without hesitation, end the dive if he stops paying attention. I know my risks and I apply a learned reduction of them before I participate. I always know the risks are there and I respect them, yet I consciously exchange the risk for the benefit.

The same is true with living in an undersea colony. There are clear risks. However, it does not take an actuarial genius to understand that by being in an undersea colony I am statistically much safer than if I lived ashore and drove an automobile around each day. So I have exchanged one risk for a lesser risk and I understand them all. In the undersea colony there will be threats just as there are in any family home ashore. Some of the threats are the same, and some are not.

Catastrophic Flooding

Of all the primal, latent and not so latent fears about living in an undersea colony, there is perhaps none greater than catastrophic flooding. An indelible image is painted in the minds of most people that depicts a massive wall of frothing, cold, green seawater gushing down the passageways of an undersea colony and snuffing out the life of everyone inside. But in the history of undersea habitation, there is not a single record of any habitat ever catastrophically flooding with an aquanaut inside; not even one. Out of the documented nearly 70 undersea habitats ever built and launched, not one has ever experienced this fate. Not one aquanaut has ever experienced a single moment of catastrophic flooding. As a mater of fact and of conscious design, most habitats constructed do not even have a door to close – their moonpools are always open to the sea! With many decades of aquanaut experiences, catastrophic flooding has not occurred with any of them.

The world's longest serving habitat, the science habitat *MarineLab*, has been on the seafloor with an open moonpool for more than 22 years and has never even gotten its deck wet, save for the many dripping aquanauts that have tread upon it. Just adjacent to *MarineLab*, the *Jules' Undersea Lodge* has been operating for more than two decades and, again, there has never even been a single misplaced drop of ocean water to report. As far as I can tell in my records, there has never even been a close call in any habitat!

That is not to say that habitats will never flood. Anything is possible. And yet to carry about this fear as the primary objection to undersea colonization has no relation to real data or experience. This apprehension is probably linked in some extent to the various movies that depict trapped passengers in sinking ships or submarines. Alarm is generated by the horrific contemplation coupled to these powerful images of being trapped and drowning in a deep ocean. It is a natural human phobia that is common to just about everyone. Such imagery has not helped alleviate the fear as a rational person thinks about living undersea.

I believe the reason why no habitat has ever catastrophically flooded is due to several factors. One: unlike the threat to military submarines and ships, no naval flotilla is out there trying to torpedo, depth charge or ram and sink manned habitats on purpose. Two: manned habitats, by definition, are fixed structures that do not move about and therefore do not run into other objects by accident. Three: a fixed undersea habitat structure is a relatively simple device and potential structural failure points are comparatively few, well planned for and engineered around.

However, just for academic discussion, let us examine the various what-ifs:

If a hull is breached or a system fails, the colony may start taking on water in an uncontrolled fashion. In order to manage this situation, emergency breathing devices should be staged throughout the colony so that colonists may have an immediate source of air during the emergency. Much like ship design, each compartment must have an air tight door that can be closed during a flooding event so that the leaking compartment can be isolated and the entire colony not flood with it.

Because of the possibility of flooding, each individual compartment must be fitted with a bottom hatch to allow for rescue in the event of a flood. For example, if the pressure hull's integrity is breached, an emergency blow-down system is triggered from the colony's command center. If the hull breach is in some location lower than the very top of the affected compartment, it is isolated by the water tight doors and blown down with pressurized air. This will create a bubble of air at the top of the cylinder for trapped colonists to breathe until they can be rescued. Raised 'air bubble' areas in each compartment specifically intended for such use may be incorporated into future colony designs.

While flooding is serious, it is not as serious as a fire since flooding can be isolated and colonists should have time and trapped air to breathe until they are rescued. With the previously mentioned system, there is never the possibility that some colonists will be isolated with no way out! The worst case scenario is that a compartment will be breached directly on the top. Yet with appropriately designed shielding for topside impacts, this possibility is minimized by careful design planning for these events in advance.

I must emphasize that the habitat structure is a relatively simple system. With proper forethought, prudent design and planning, catastrophic flooding, while it may present the greatest fear, may ironically actually be the least of an aquanaut's worries.

Fire

While it is true that catastrophic flooding is the most feared of all threats by the surface novice, to the experienced aquanaut, fire is unquestionably the greatest of all fears and actual threats to the undersea colony. Although you may intuitively suppose that the ultimate and most dreaded of all threats would be drowning, the history of living and working undersea does not bear that out. And the idea of a raging fire underwater is also something of a surprising idea. But if you dispassionately analyze the hazard probabilities, a fire in a habitat or colony would certainly top the list as the most dangerous of all possible threats. Fire is a serious hazard because even a small fire consumes oxygen and produces toxic smoke. Fire is also a hazard in an undersea colony because there may be nowhere to retreat, and removing the smoke from a colony is problematic.

When astronaut Jerry Linenger flew a space mission on the aging Russian *Mir* Space Station, they experienced a fire onboard. It took 14 minutes to extinguish the fire during which time the station filled with thick, acrid smoke. Their procedures required that Linenger and the other astronauts don gas masks, but they had to wear these masks nearly an hour while the filters worked to rid the air of the smoke. Linenger's first respirator failed, several fire extinguishers failed, and Linenger's lungs felt as if they were on fire. At the height of the catastrophe, he said a silent prayer of goodbye to his pregnant wife and son. Fortunately, the fire was eventually controlled and he survived the ordeal.

If a fire in an undersea colony was small and quickly contained, the exact same scenario would unfold. If the fire were any larger, it would begin to consume the available oxygen and the crew would lose consciousness, inhale the smoke and likely die before they could act.

The three best defenses against fire are:

1. Fire Prevention: Head a potential fire off before it starts. Use incombustible materials whenever possible. Officially inspect at least once per day for possible fire causing agents and eliminate them. Everyone in the colony should responsibly act as fire inspectors.

2. Fire Control: Plan for fire. Install fire and smoke detectors in appropriate places. Wire these to a community wide warning system. Have a community communications system to let colonists know what is going on and what actions to take during any emergency. Train for fire escape and hold scheduled drills. Train everyone to don emergency gear and be able to intelligently use fire extinguishing equipment.

3. Viable Retreat: Have a route of escape. Ensure that colonists have a 'life boat' to retreat to in the event of a colony-wide fire. No colony should ever be built that does not have an installed 'life boat'. (In an undersea colony, a 'life boat' is a compartment or compartments that can fit every colonist and sustain them until evacuation, if that is necessary.)

Hurricanes, Typhoons and Storms

For colonies in hurricane or typhoon prone waters, these are the next biggest threat. As we have discussed in previous chapters, these storms come with a mighty punch of hydrodynamic energy. It is very probable that colonists in the path of a storm will be evacuated. In this event, the greatest threat to the colonists will be their attitude. If they see themselves as 'permanent' colonists with no intention of ever returning to the surface, they may resist evacuation.

As Mission Commander, I would require that all possible crew members submit to a psychological evaluation to pre-determine any such mindsets. 'Permanence' as a resident of the sea in no way means that they will never again stick their heads above the surface. It means the Aquatican views the undersea dominion as their permanent home. It does not mean they will never visit relatives, take vacations or in many cases take a shore based job and return home to Aquatica each day. And it most certainly does not mean they cannot leave in the face of a hurricane! It is no more of a 'shame' for an Aquatican to evacuate to safety in a hurricane threat than any other land based citizen. A well placed hurricane or typhoon has just as much potential to leave an undersea colony in shambles as it does any other city.

After fire, flooding and storms, there are many hazards that remain – but no more than are faced by anyone who lives in an average land based residential home. If you view the nightly news long enough, the possibilities for personal harm are endless; from cars that crash through living rooms and bed rooms on a somewhat regular basis, to falling trees, parts falling off of aircraft and the occasional meteor or lightening strike. While the undersea colonist can strike the meteor worry off their list, they have a few others to be concerned about.

Dropping Anchors and Collisions

One prevailing worry that has always struck me as somewhat odd is the concern over falling anchors. In order for a colony to be struck with a falling anchor, a ship would have to be situated directly overhead, stopped and the captain order the release of the anchor on purpose. The odds against that happening given the overhead caution and warning system connected to the S3 are vanishingly small. The release of a shipboard anchor is not a one man operation and, with the assembled crew required to accomplish that nontrivial feat, it would be nearly impossible that they would not see they were in a well lit and defined area of the ocean. Further, they would have to fail to notice they were sitting in an ocean filled with flashing beacons and an undersea command center duty officer screaming his head off on their guard channel. The colonists can rest easy in that the meteor strike is a far more likely concern.

In this regard, surface ship hazards are fairly small in general. Most colonies will be considerably deeper than the draft of even the largest aircraft carrier so that these behemoths of the sea could easily and safely pass directly overhead.

One hazard that has occupied my thoughts is an undersea collision with a submarine or exposure to their sonar. Undersea colonies will purposefully not be located in any 'nominal' submarine tracks, but submarines have the free run of any reasonable ocean depth in which most colonies will be located. Obviously, the colonies will be clearly marked on all navigational charts and as soon as they are in place, they will be immediately annotated in the daily 'Notice to Mariners' and then hopefully penciled on the charts by the navigator.

But submarines are most famous for their conspicuous lack of windows and if they fail to update their charts or the navigator is not paying attention, it could be a very bad day for everyone. The seas of the world are full of submarines from many nations and to prevent undersea collisions, everybody has to be paying attention. It is likely that all undersea colonies will have some form of acoustic warning device so that if a submarine passes within their range, they can activate it immediately. This requires that the submariners of the world know what that noise means and that it therefore warns them away instead of actually attracting them!

The other hazard is exposure to the submarine's powerful sonar. Inside an undersea colony, if a submarine is located 'feet' away and pings the colony directly, the acoustic energy could cause damage to the structure or harm the hearing of those inside. However, the undersea colony will be the noisiest thing in the oceans and submarines will certainly hear it miles away. Hopefully they will understand the source of that noise by looking on their corrected charts before they rush over to see what it is! Of all the nameless hazards, submarines worry me the most.

This chapter represents the short list of hazards, but they are probably the most infamous ones. The hazards that are truly the most dangerous ones are, of course, the ones you do not know about and hence are not planned for.

But of all the threats to the undersea colony, there is none greater than the threat of human politics where logic or intelligence may not necessarily be a part of the equation.

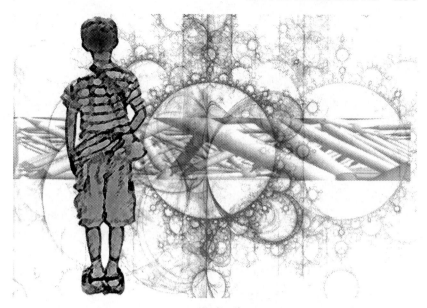

POLITICS

"With malice toward none; with charity for all;
with firmness in the right,
as God gives us to see the right,
let us strive on to finish the work we are in."
Abraham Lincoln

I n 1996, Claudia and I were invited to Biosphere II near Oracle, Arizona, to tour the facility and offer some insights into what may have been an apparent biome shift occurring in and around the Biosphere II's ocean. The Biosphere facility was famed for having hosted a rare, biological experiment. It was the first time humans had created what was considered to be as close as possible to a natural biosphere. Inside the 3.15 acre glassed sphere structure was a mini-world within a world that attempted to duplicate a natural biosphereic cycle with a crew of eight humans locked inside for two years. Much was learned. One of the key lessons learned was how politics, when integrated with science, can become a very ugly thing and devastating to the good order and discipline of the expedition itself.

When we arrived in Oracle, I was determined to keep an open mind and filter all the political trash out. I wanted to see the actual science of Biosphere II and was not at all interested in rumors. Long before we arrived, I had already heard them all. But I had not come to partake in more rumor mongering; I wanted to see what actual data they might share.

Unfortunately, long before I slipped beneath the waters of Biosphere II's chilly artificial ocean, I had already heard about the fights. There were arguments between the corporate structure that represented Biosphere II and just about everyone else. On the part of all parties, there were lawyers involved, hearings, infringements, insults aplenty, real lawsuits, threats and even physical attacks. Finally, just before we arrived, Columbia University had purchased the facility and some semblance of sanity had settled in. It was also very fortunate that a few of the original scientific staff still remained, and they were the ones I wanted to talk to. Our conversations were more than fascinating, but that is for another book.

Who was at fault for the very unfortunate state of affairs just before, during and after the two year experiment? I very much wish to avoid entanglement in a lawsuit aimed at me, but I can truthfully state: just about everybody on all sides. The level of jealousy within the scientific community in general was to me professionally embarrassing. Added to that was the digging into the past history of the Biospherians to find any dirt that could be found and then leaking what was discovered to the media. I got sick of hearing about the $200 million price tag and "all that could do for 'real' science". One critic called it, "New Age drivel masquerading as science".[a] The disaster was truly sad.

By the time the experiment was over and the unprecedented data was ready for analysis, no one seemed to want to go anywhere near the place. It had been so maligned and infected that anything of value that had been discovered was hopelessly tainted by the tragic mess that had absolutely nothing whatsoever to do with the data they were actually recording. It was, from start to finish, a public relations and political nightmare and a scientific tragedy.

Biosphere II seemed to make every public relations error possible. It was spun to the media as frontier science, but oddly, it was not executed as a scientific project at all but a kind of daily performance, featuring pre-staged acts infused throughout. It was billed as a two year sealed 'mini-earth' that would invariably prove the concept of Biospherics and achieve its own one-normal earth balance just as they predicted – instead of a truly hypothesis based scientific approach of, "we'll fine tune this hypothesis for two years and see what happens". Therefore, when they almost unavoidably had to add oxygen to keep the crew from slipping into anoxia and chose not to make an announcement of the fact, the media headlined them as 'cheating' and the experiment as a 'failure'. Everyone was at fault and neither the participants nor the media had any clue how it actually all fit into 'real science'.

BIOSPHERE II

Paradoxically, the project actually achieved a brilliant scientific success that was totally lost to the media circus that eventually engulfed and consumed it. But they were not alone. It was equally lost to a scornful scientific community that had so many reasons – including professional jealousy – to despise the people involved that it finally refused to even look and see what they had actually discovered. But some of us managed to visit and see what Biosphere had really discovered: a marvelous unearthing of unexpected and important information that will not be fully appreciated for many decades to come.

Do I say this because I am 'friends' with the Biosphere II crew? No! I only met one of them in passing. I am friends with one scientist who was on the staff there, but that in no way affects my judgment of what I witnessed with my own eyes.

The lesson here for everyone could comprise a graduate level course in how not to run a project. And as for the media, the lesson is how not to report on science and just what happens when journalism becomes a destructive force instead of a constructive power in any culture. As far as science is concerned, we must learn the meaning of 'dispassionate analysis' and purposefully jump off the media's train. Hopefully, as we prepare to permanently occupy the oceans, those lessons will not be lost on the future crews and project directors as well.

What lessons do we as Aquaticans learn from Biosphere? The first and most important is in the open communication of terms. And the most important term for the Aquatican is: 'permanence'. In this application, permanence means a 'permanent address' and nothing more. It certainly does not mean that the Aquatican will never stick his head up out of the ocean nor does it mean we will not make occasional or even frequent trips ashore. Many Aquaticans will actually go to their jobs ashore each day! And if a hurricane approaches, all of us will travel to the shore for safety. But that does not make us failures or cheaters. It makes us people with a permanent address undersea who have total freedom of movement, just like anyone else.

Does it make the reporter a 'cheater' or a 'failure' because his permanent address is located in some residential neighborhood, but he works in another part of town or takes his car to the movies once in awhile? Of course not! Does it make the scientist a failure because he drops work in his laboratory to evacuate for an approaching storm? Such a suggestion would be labeled as patent stupidity. But beware of the lessons of Biosphere II when dealing with critical and often badly informed people.

I am certainly not a prophet, but as we examine the many experiences of Biosphere II, it is possible to discover from their lessons-learned how to wisely anticipate the political roadblocks to establishing man's permanent presence undersea. Yet, in the end, Biosphere II's greatest enemy and ultimate roadblock was itself.

Unfortunately, they attempted to 'play' the media. As a former journalist, I can assure anyone who wants to listen that a journalist has a finely tuned sense of being 'played' and being used as a publicity agent. It causes instant resentment and can turn an otherwise friendly reporter into a critical one whose first impulse is to look for problems. After all, in this circumstance, it is the only way the journalist can retain their professional integrity. In this situation, they really have no other choice but to be critical. If you want a story, they'll give you a story – but you may not like what you read. There is only one way to deal with any professional journalist: absolute, no holds barred *honesty*. This they understand and respect.

Biosphere II then attempted to exploit science and scientists to their advantage. This usually occurs by stacking willing scientists on the 'board of advisors', followed by speaking in flawed 'scienglish' to the media which actual degreed scientists pick up on immediately. The response by the scientific community is to typically create as much distance between themselves and the project as is humanly possible. Yet when cornered and asked, they will not have very many nice things to say. In the case of Biosphere II, the scientific community was at first clearly communicating interest and a willingness to help, but as the secrecy and scienglish began to form the image of the project, the scientific community backed away – and for very good reason.

All of these examples are lessons to be learned for the willing. The bottom line in project politics is: do not 'play' anyone for your advantage, ever. Let them come to you on their own terms, and then speak the whole truth. If no one comes to you, then be happy – there will be less resistance. Artificial 'fame' is fleeting and dangerous – leave it alone. Focus on the long term goals and allow the popular culture to do whatever it wishes – true science has never blended very well with pop culture. If you do not know, just say so; everyone respects honesty. Define your terms clearly. Never, ever speak scienglish to anyone.

Use your own press agent for your own press releases and never abuse a professional journalist – it will never work to your advantage. Be particularly kind and respectful to everyone, even if, and especially if, they are abusive to you. Never give them the fight they are spoiling for because you do not have the time. If someone else has a major problem with your project, do not allow them to sidetrack something that never needed any defense to begin with.   Allow your attorneys to work the corporate paperwork and not busy themselves with suing people who merely say unkind things.

But the most important lesson is this: keep your eyes on the goal and off of yourself. The minute you focus on yourself and not the worthy goal, the project is over. These are the lessons I took from Biosphere II and hope to pass on to those who dare great things in new frontiers.

Unfortunately, the politics do not begin and end with project management! In establishing the ocean as a frontier for human expansion, there will be many opponents; it is inevitable. Still, we must press on toward the worthy goal. As Samuel Johnson has so aptly acknowledged, "Nothing at all will be attempted if all possible objections must first be overcome."

One objection I found most surprising comes from a very unexpected source: space advocacy. I have heard many complaints from their quarter that the undersea colonization will somehow 'detract' from space colonization. Or, they have protested that conditions will be horrible 'down there'. Both of these sentiments are off-the-scale strange. If you compare the dollars spent for space to those spent on human exploration of the oceans, it is not even on the same scale! And conditions undersea will almost entirely mimic conditions in any space colony.

But perhaps the most vocal objections come from environmentalists. I find this particularly heartbreaking for I, too, am an environmentalist. Because of that, one of our main objectives is a new, powerful method of monitoring and protecting the seas. Some would try and stop us in the very name of the process by which the seas will be protected. It must be the ultimate irony! Yet, before I despair totally, I have to consider just who it is that protests.

I have no use for radical environmentalism. Philosophically, it is without question much more of a religion than a science. Radical environmentalists use the word 'science' as a validation of their philosophy, but real science is actually not a part of the process at all. I also despise making up data, back fitting data or mining data specifically to fit a social or political point of view. That is nothing more than blatant dishonesty and represents a wholesale misapplication, misunderstanding and purposeful manipulation of the scientific method. But worst of all is the mixing of politics with science – that is yet another form of corruption whose unhidden objective is to force free people to do things, or stop them from doing things, that are based on opinions and not reality or real data. It is nothing more than the exploitation of vague fears expressed as philosophic nonsense.

Yet, there is also such a thing as true scientific environmentalism based on actual science and real data. It is as powerful and as important as any other human objective. Sadly, true environmentalism has become so clouded with the religion of radical environmentalism that often it is nearly impossible to tell them apart – and many times their issues have become hopelessly entangled.

On the other side, I also equally despise methods and politics that sideline true environmental tribulations in the name of 'freedom'. There are deep, critical problems in our seas that depend on sound legislation to keep the living seas and their life from being destroyed by man. Many of these issues are being hidden behind the name of 'freedom.' That is nothing more than the dishonest political expression of special interest groups who make money from raping the oceans and hiding behind their corporate lawyers and pocket-politicians.

Here is one good example of many: according to a 2007 report, the world's oceans are now critically over-fished. Industrial fishing has wiped out 90 percent of large predatory fish. The United Nations reports a third of all fish stocks eaten by mankind are in the same peril. And much of this activity occurs by business concerns that are poaching or ignoring international law.[b] Except for sparse, hit and miss reporting of the annual catch by some nations, we would not even be aware of this!

There are no eyes in the world's oceans to tell the story and all of mankind must instead depend on irregular, meager and totally inconsistent coverage of annual fishery data to know what is actually going on in our planetary biome. If this situation were occurring on the land areas involving cattle or swine, the world's nations would dispatch teams of thousands of watchers, observers and eyewitnesses to hotspots all over the globe. It would be the only rational and responsible thing to do. And yet, in the largest and most critical of the world's biological systems, which provides one of mankind's critical sources of protein, we shamefully do nothing, sending not a single team to investigate this impending planetary disaster for the long haul.

Yet another of many ongoing disasters now developing in the vast, three dimensional ocean sphere is the wholesale dumping of every conceivable human created waste at sea. It is a process that is so pervasive, that as we view the oceans, it has become apparent even on that two dimensional interface. Here is one tragic view of just how widespread it has become from Charles Moore, writing in *Natural History* about his experience after racing his boat across the North Pacific in 2003:

"Throughout the race our strategy, like that of every other boat in the race, had been mainly to avoid the North Pacific subtropical gyre – the great high-pressure system in the central Pacific Ocean that, most of the time, is centered just north of the racecourse and halfway between Hawaii and the mainland. But after our success with the race we were feeling mellow and unhurried, and our vessel was equipped with auxiliary twin diesels and carried an extra supply of fuel. So on the way back to our home port in Long Beach, California, we decided to take a shortcut through the gyre, which few seafarers ever cross. Fishermen shun it because its waters lack the nutrients to support an abundant catch. Sailors dodge it because it lacks the wind to propel their sailboats.

I often struggle to find words that will communicate the vastness of the Pacific Ocean to people who have never been to sea.

Day after day, *Alguita* was the only vehicle on a highway without landmarks, stretching from horizon to horizon. Yet as I gazed from the deck at the surface of what ought to have been a pristine ocean, I was confronted, as far as the eye could see, with the sight of plastic.

It seemed unbelievable, but I never found a clear spot. In the week it took to cross the subtropical high, no matter what time of day I looked, plastic debris was floating everywhere: bottles, bottle caps, wrappers, fragments. Months later, after I discussed what I had seen with the oceanographer Curtis Ebbesmeyer, perhaps the world's leading expert on flotsam, he began referring to the area as the 'eastern garbage patch.' But 'patch' doesn't begin to convey the reality. Ebbesmeyer has estimated that the area, nearly covered with floating plastic debris, is roughly the size of Texas."[c]

Here is an eyewitness view from the surface. But if all of this is apparent on the surface of the massive gyre, what was going on beneath the waters? Moore decided to go back and investigate the mess floating over the area of the 'Eastern Garbage Patch' and began collecting debris and examining it. Here was his analysis:

"In 2001, in the *Marine Pollution Bulletin*, we published the results of our survey and the analysis we had made of the debris, reporting, among other things, that there are six pounds of plastic floating in the North Pacific subtropical gyre for every pound of naturally occurring zooplankton."

While these realities are staggering and sickening, what they did not look for was the dissolved chemicals that came from all that waste because they were focused on plastics and floating debris. What they witnessed was only a fraction of the damage actually done. Most of it is unseen and not even viewable from the surface. And no one could compare the ongoing effect on the zooplankton to any past data because in our irresponsibility, we have not collected it. Major and significant questions still go unanswered: what effect does it have in the ocean layers? What is happening to the zooplankton over time? What is the effect on the ocean life of all sizes and species?

If we established undersea stations today fixed within the gyre, we could monitor data continuously as the gyre spins by and thereby collect a complete data set on hundreds of variables on this Texas sized area as it literally orbits around an undersea station. But the information we collect will be the first data set forth for future ones to reference. Unfortunately, we have lost all prior data to date because we are not there to collect it. And this is only one critical ocean hotspot that goes unmonitored.

Tragically, in our environmental zeal, we have blocked the way of the emergency responders to a homicide in progress because of a vague fear that one of them may be a bad guy! Of all the species on earth, only humans can become so befuddled by virtuous enthusiasm that it interferes with cogent reasoning and logic, and our own intelligence often leads us down a path of self destruction.

How do we respond to these problems? Some say, "If we can't see it, then it can't harm us." But more realistically, we need to be in the oceans today – right now – with intelligent human eyes sent for the specific purpose of permanent environmental monitoring. The military is there in submarines, but it is not their mission to monitor environmental change and they certainly do not! The fisheries are there on the surface, but it is their objective to take their haul, never to monitor, and they rarely collect data on the environment. Merchant ships ply the surface to load and unload cargo, but never to collect data. Cruise ships take their passengers out to party, but not one of them is engaged in data collection. The occasional scientific vessel drops probes into the water once and awhile, but their purpose is narrow and focused on their relatively small and individually funded objectives. We are blind there and most of our planet is totally off anyone's radar.

We need permanent eyes in the ocean and we needed them there yesterday. It may be one of the most urgent objectives in all of humanity. We are now totally sightless to three quarters of our planet and there is no sense of urgency to correct that problem while we play political games with one another and argue about the need for real data. We know more about the surfaces of most planets in our solar system than we know about our own. And yet our billions are spent on those distant bodies while not one single human colony exists to monitor our seas. It is more than an enigma – it is the expression of wholesale irresponsibility on our part, and our children and theirs will pay the price for our self-indulgent capriciousness.

How is this problem solved? Just like any other: with good science; with well informed political support; with cooperation between special interests; with fair media reporting; and with project managers who are more interested in the objective than their own fame and making a name for themselves.

As a planet, we are quickly approaching a win-or-lose time of crisis. We are staged to lose control of our principal biome: the seas – a vast region that supports each of us with everything from food and oxygen to planetary heat exchange and carbon dioxide sequestration. We are losing the fine balance that supports us all. And it is not just biological. There are also physio-chemical processes that may also be swinging far out of their natural, cyclic balance.

As a species, we face a critical, imminent problem that will dwarf the issue of climate change in its implications and its direct effects on humankind. The key debate today regarding global climate change revolves around its chief cause: is it a cyclic change we are witnessing or is it strictly anthropogenic? An honest person, when faced with the complete data set, devoid of the fashionable religious zeal, cannot truthfully say with absolute certainty.

But in the case of the oceans, there is no debate. The chief culprit contaminating and killing our ocean environment is man, period. No balanced, natural cycle could ever provoke such devastating havoc. Now we are faced with a question that has larger implications than mere global warming. When the life blood of the entire planet is being chronically poisoned and mined for its living quintessence, then what do we intend to do about it? That question is absolutely dependent on going undersea as soon as it can be arranged and finding out what is actually going on. Astonishingly, the first comprehensive data set has yet to be collected!

The current scenario does not have to end in planetary disaster. There may still be a chance to rescue our oceans. There are people who are ready and willing to make it happen. But we certainly have no time remaining to play politics and self destructive games.

And yet, while we wrestle with politics, we also wrestle with other possible legal entanglements as well.

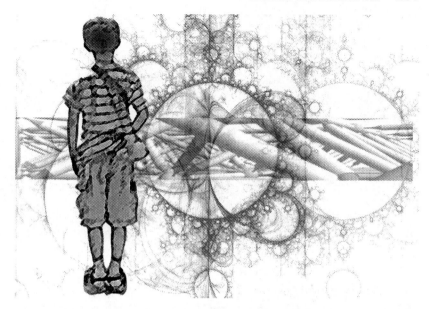

LEGAL ENTANGLEMENTS

"It is not what a lawyer tells me I may do; but what humanity,
reason, and justice tell me I ought to do."
Edmund Burke

A s the idea of walking boldly into a new frontier first begins to dawn on our culture as a truly serious notion, I am quite certain these next four questions will pop up after only a few minutes discussion with any reasonable person. And well they should. There are literally more lawyers interested in any undersea project than bull sharks cruising just off of Cocoa Beach. And yet, as essential as it may be, this discussion is unpleasantly unlike science and engineering. Here, the gray areas rule and there is no 'right answer'. As a scientist and engineer, it is all very maddening to me.

The great Danish scientist and mathematician Piet Hein said of this discussion, "Just beyond perception's reach I think sometimes I see, that life is two locked boxes, each containing the other's key." The English translation of that very sentiment is more direct: "You can't get there from here!" Here are four questions that seem to illustrate these very often frustrating sentiments:

Who said we can go do this in the first place?

The United States Constitution. This is, after all, America. Here, we can do just about any crazy thing we like, just as long as the public is not endangered. Yet, the politicians of the world lay awake all night dreaming up ways to 'protect' the general public (but unfortunately never from themselves.) I am not making *any* of the following examples up! All of these are right out of the law books:

Even though I am a citizen of a free capitalistic republic, by law in Texas, I may not read any book of the Encyclopedia Britannica because it contains the formula for making beer.

Here in my home state of Florida, unmarried women who parachute on Sunday's will be jailed, while in my birth state of Oklahoma, people who make 'ugly faces' at dogs may be fined and/or jailed.

And finally, the Massachusetts township of North Andover prohibits its citizens from carrying 'space guns'.

So, as you can clearly see from these examples, there are some things I may not specifically do if they endanger the safety and welfare of others around me. I am so happy that my elected representatives are so dutifully engaged and I *feel so much safer now...*

But, it would seem to be a pretty safe bet that if we intend to migrate out to where the population is, and has always been, zero, that the public is really, really safe!

"Perhaps..." murmurs the local law society while rubbing their chins. But the fact is, if we all decide to go do this, there are no laws (yet) that can stop us!

And just to make things really interesting, if the US did pass any laws, would they apply out beyond the three mile international boundary, stop at the 12 mile limit or extend all the way out to the 200 mile exclusive economic zone? What I can assure you is this: we are now definitely into some gray legal areas that have never been tested before!

If we build undersea habitats, submarines and chambers, who is it that gives us permission to operate them?

That depends. (Don't you just hate it when people answer a question like that?)

It depends on whether you take people with you, and who those people are, and if you charge money to take them along. All of those 'what if's' label the members of the party differently. Are they 'company employees', 'corporate volunteers', 'visitors', 'passengers' or the 'general public'? Each one of those classes changes the application of liability somewhat.

But in the end, no one really gives us permission to operate underwater. The question is not necessarily permission, since 'research vessels' are not necessarily licensed by anyone. The question to the organizing group is not permission or licensing, but liability. Liability, after a very bad day, can range anywhere from personal criminal negligence to corporate negligence to merely an accident. Liability is sometimes termed 'exposure' which means the people who organize the events are 'exposed' to lawsuits or even personal criminal charges if things go really wrong, which leads to the next question:

If someone gets hurt or dies in our structure, who gets sued, and have we broken any laws?

That depends. (Sighhh…)

It depends on what happened and how it happened. It is possible that the leadership may be charged with a crime, depending on what happened during the accident or what led to its cause. It is more likely that a civil suit would be brought, to recover damages, in a courtroom setting. It is also possible that the 'legal authorities' would order a Board of Inquiry, depending on the events and circumstances.

There are precious few laws and regulations governing living and working under the sea. Most laws are made after some unfortunate accident or tragedy, and these (reasonable) laws are mostly written to prevent such events from happening again.

How do we protect ourselves from liability in a tort crazy world?

In a word: insurance, of course.

If you look at most habitat programs, some form of insurance is in effect. Government habitat programs are all 'self-insured' – which means the United States government is willing to accept its 'exposure' and 'cover it' in the even of an accident or tragedy. Most SCUBA dive shops have insurance policies to cover their dive activities.

It is possible to obtain insurance, but the insurance underwriters are not fools! They want maximum assurance that they are not covering an accident waiting to happen, so they rely on other agencies to provide results of inspections of the undersea systems to assure them they are as safe as possible before they will extend the coverage.

For example, there is an agency called the American Bureau of Shipping (ABS) that has published a document called "Underwater Vehicles, Systems and Hyperbaric Facilities" that covers all parts of these systems in excruciating detail and establishes standards for their construction and operation. Further, it feathers in quite nicely with another organization and document written by the American Society of Mechanical Engineers (ASME) called, "Safety Standard for Pressure Vessels for Human Occupancy". If an organization that operates underwater facilities wants any insurance at all, they must comply with these documents or the insurers could decline to offer coverage.

But, since an ambient pressure habitat is not a pressure vessel, then what? Just more gray…

The only problem with all this is that the undersea world is going to end up looking a lot different than we now think. For example, most of the design specs in these documents are oriented for steel or metal vessels. It is very possible that future vessels will be concrete, acrylic, fiberglass or even accreted minerals from seawater! One cannot blame ASME or ABS – they have never encountered these materials before in undersea designs. I am certain they will keep up, but the initial creations may well not be covered by any insurance because their technology will advance far faster than these agencies can document them!

We now exist in a miasma not unlike the haze Columbus and his crews sailed through to the new world just beyond their sight. His ships and ours are definitely sailing into the unknown more rapidly than we can map the way before us.

But as we work to make permanence undersea a reality, there is a certain, palatable tension that exists between what we *must* do and what we *should* do. No one wants legislated requirements, that are usually poor or impossible technological fits to any new expertise, and are always hideously expensive. Indeed, the tourist submarine industry even now complains that inspections of their vessels often cost more than the vessels themselves! And yet, if I and my family are riding in one, I definitely want that process to be as good as they can possibly make it! Before I get in, I want to make sure that somebody else has been asking questions, besides the man taking my money.

So, the issue for all of us is how do we get the assurance that the responsibility of the ones that are inviting us to go live undersea is as serious as it needs to be and that they have their eye on the nuts and bolts and not just on their bank account? And for those of us who intend to make it happen, how do we assure ourselves and others that we are up to the level of the task?

In our culture, there are two avenues to approach these serious and vital questions, as onerous as they may be. We can have the politicians tell us how to do it or we can police ourselves by implementing the ABS and the ASME standards in our undersea equipment wherever possible. And what is it that ultimately drives us to take the extra steps? The ever present threat of someone else's lawyer, of course.

Believe it or not, the entire process is one of checks and balances and it sets up a very powerful equilibrium for everyone! Here is how a colleague of mine, professional diver Rob Bryan, described this system of balances:

"The tort system is as great an equalizer as the firearm was. Giant 'for profit only' corporations can no longer use up and throw away people like they did in previous era's and as some cruise ship lines still do. The alternatives are government regulation or anarchy. I have only small guns and don't like the way governments change things depending on who's paying them to, so I like this system best.

"I'm not a Friedman free market advocate. It's a great idea but it doesn't now, and has never, existed on this planet. I'm not a socialist, either. I like capitalism. But capitalism without controls is anarchy – the biggest gun or the best organized crime syndicate rules. Capitalism needs regulation. In the absence of government regulation, the tort system is the best substitute we've come up with. Disdain for liability responsibility (as a court would see it) increases the exposure. My view is that liability management, quality control, and safety management are all part of the same thing.

"Putting people on and under the sea is a serious responsibility. I don't mean to sound tutorial, if I do, but, on more than one occasion, I have literally packed my bag and 'drug up' (quit, in the vernacular) rather than make a dangerous dive for someone else's profit. I was very lucky to have learned the potential consequences at someone else's expense in a dive school accident. Last I heard he was walking and talking but never dove again. In the 80's a dive company in the Gulf could kill 3 or 4 divers before insurance became a problem. Not so today, and liability is the reason, not government regulation."

Finally, unlike most lawyer jokes which rarely have a good ending for my attorney friends (yes, I admit it, I do actually count a few of them among my friends ...), the attorney truly deserves great and abiding respect as a formidable force of equilibrium in our world that extends even into Aquatica. And just as essential as the attorney, is the engineer who is driven to design for safety first and the expeditions leader who knows that there will be a tomorrow only if he respects and is willing to pay the full cost of a quality process.

In the end, when we submerge our ships, vessels, habitats and boats beneath the surface interface, we musk never forget who it is that rides with us and who it is that we are designing for in the first place: Aunt Miriam and her kids – the families that will make Aquatica what it will become. All of them deserve nothing less than our best and our all. And that is one sentiment that all parties can completely agree upon!

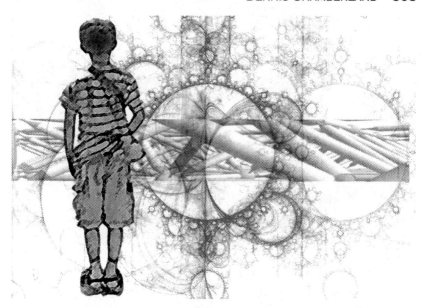

Families in the Sea

"You guys have an actual undersea family here!"
Writer and Director James Cameron to Claudia Chamberland

I f the chapter on Environmental Stewardship is the 'most important' chapter in this book, it is this chapter that is my favorite and the one closest to my heart. It is this theme around which the whole of the book is focused. Families in the sea represent the true future of human permanence in the oceans. Every other idea and every other model are all just precursors to this one. It is also very special to me because, for a time, my own family lived and worked together in the sea.

Human expansion has always proceeded in phases. It begins with the first humans to go and check the new regions out – to see what is there. The second phase is the adventurers that go and see what they can do with the new place. They are historically followed closely by families who come to settle and to expand. You can send as many explorers and adventurers as you please, but after they are dead it will be as though they never came at all. Without families there is no future and, ultimately, no hope for expansion. It has been said that the hand that rocks the cradle rules the world. This is especially true for the world undersea.

Having established this crucial viewpoint, allow me now to make a bold claim that will invariably cause much controversy: the undersea dominion of Aquatica is now ready for its first families. Man has gone there, sent our explorers and adventurers, learned from their mistakes, has mapped out the viable technologies and we are today now ready for full occupation.

It was not too long ago that when we made statements like that we invariably considered 'the women and children'. In the 21st century, half of that equation has radically changed – and for the better. We now recognize that women are certainly just as capable and qualified as explorers and hardy adventurers as any men. To make any distinction between them based merely on their sex is unethical and it is wrong. I know this very well – I am married to one!

Claudia Chamberland not only keeps up with me and my ideas and explorations, she has been by my side every step of the way. I would *never* consider going on any mission without her, period. She is not just my partner in life, but in every phase and part of my life. If I dared go explore any part of the oceans or space, she would be standing beside me, diving with me and holding my rope as I hold hers. Whatever adventure or exploration I am involved in, including the writing of this book, she is as deeply involved in it as I am. She has aptly and awesomely modeled to me that leaving women out of this equation is not only wrong, it is a dim-witted relic of a stubborn, pig-headed age that has mercifully passed. Thankfully, there is no more 'women and children' ethos – leaving only the children to consider.

On the following pages are photos of three of our sons: Eric Milton, Brett William and Eric Alexander . As I stated previously in this book, our children accompanied Claudia and me on our expeditions because they were home schooled. We encountered all of these experiences as a family. Many of our children are card carrying certified SCUBA divers and Eric Milton is a certified aquanaut. It is possible, although I have no way of confirming this, that the Chamberland family may be the first full aquanaut family of parents and children to live and work in the sea as researchers and scientists other than resort visitors.

AQUANAUT ERIC M. SITTING ON THE HABITAT DURING DEPLOYMENT

But long before the habitat pictured here in Key Largo was launched, the Chamberland family had built and experimented with a habitat in another Florida location. Long before this photo was taken, the Chamberland family had established an undersea location and built habitats as a family experience. We were ready for Aquatica long ago. Had a colony been established, I most certainly would have moved my family there! And I would move today.

Yet, the sensitivity is understood. Before the 21st century, in a much less enlightened age, women would have been bunched with the children in a kind of general fear category labeled, "It's too dangerous. It's too early to pack along the women and children." As was written before and is now repeated for emphasis, categorizing the women with the children is accurately characterized as Neanderthal thinking, and for good reason. But the children are and will always be a special case. No amount of social engineering will ever change their state. They are vulnerable and they all require special protection.

The next obvious question is: Protection from what? Certainly every parent has the obligation to protect their children from any hazard – physical or emotional – that might harm them now or in the future. But from there, it gets fuzzy. Some parents think nothing of taking their children along on rides in tourist submarines. Thousands of pre-teen children ride these submarines each year. None of these parents, as far as I am aware, have been charged with child abuse or negligence. Instead, they are joyously clinging to their children as everyone looks out the windows and has a good time.[12]

DIVER ERIC A.

Millions of children each year tour the world's aquariums, all of them featuring tanks and even open bays with millions of gallons of water that would flood and kill everyone if even a single window failed. And yet, no one seems concerned about their safety.

Tens of millions of children fly in jetliners annually with a multitude of very dangerous conditions all around them. Sadly, not all of them make it home alive. Yet the probability of dying in

---

[12] They can do this because the tourist submarines are one atmosphere boats with comfortable shirtsleeve environments. They have already successfully modeled for all of us the essential nature and power of the First Principle of Undersea Colonization.

a jetliner accident is far, far less than the automobiles we, without a second thought, strap our children into each day. Even more sadly, tens of thousands of them do not make it home each year while many others are permanently maimed.   Oddly enough, in our anomalous and rather schizophrenic social consciousness, this is considered fully acceptable.   It would be unthinkable that anyone would be charged with child abuse who allowed their children to ride with them in an automobile, and yet the odds of their injury are so high.

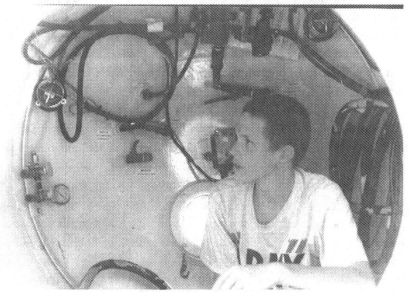

DIVER BRETT W. INSIDE SEALAB
PERSONNEL TRANSFER CHAMBER

What is the difference?   Not a single child has ever gone undersea to live permanently, so there is no social awareness of the issue and in the place of awareness, the mind almost always substitutes an unspecified fear. Before that fear will be replaced with the same ease of mind as tourist submarines, aquariums, jetliners and automobiles, the undersea colonist community will have to educate the general population about its relative safety for everyone.

The core philosophy of this book as been that the "Protection of the human is the primary objective."   The pioneers discovered that no man or woman, however strong, burly and

Arnold Schwarzenegger-ish they may be, can live in hyperbaric atmospheres with bizarre gas mixtures for long periods of time. For the last half century, pioneers and adventurers have learned the hard way. The lesson they brought back to us is that we must protect everyone in our colonies – in the oceans or in space – by ensuring everyone lives in a surface-normal environment.

When families in America assemble at home each evening, they all assemble in similar environmental conditions – one atmosphere, comfortable temperature and humidity, and protection from fire and other potential disasters. The Aquatican family will assemble in exactly the same environment under the precise same conditions in habitats that are designed for maximum safety – just like the tourist submarines and just like the aquariums, with less risk than a commercial jetliner and far, far less than the risk of injury in a family automobile.

As I pictured before, Claudia and I stand figuratively on the shorelines of Central Florida with our cat, Snickers, waiting for a place to go.[13] Sadly, there is no place to go – not off the shores of Central Florida or anywhere else. Aquatica is empty and its population remains at zero. But the picture I paint here is focused on who it is that is standing at the shore. It is a family – not an explorer, not an adventurer, not a scientist. Although we have been and will always be all of those, it is not the explorer or the adventurer who waits for a place in Aquatica – *it is my family.* And as soon as I finally get there, make the beds and hang pictures up on the walls, I will invite my grown children, their spouses and the grandkids for frequent visits.

It is my fondest desire that as readers peruse the pages of this book, that message has been communicated most clearly. Aquatica is a new frontier. While there will be new discoveries and new explorations aplenty, it is now a place for families. The explorers and the adventurers have blazed the trail. They have made their discoveries that will allow those who follow in their footsteps to understand the true nature of the environment and the

---

[13] Our six children, five boys and a girl, have grown and left home now, starting families and lives of their own. In order of age they are: Christopher Donald, Eric Milton, Brett William, Eric Alexander, Katy Ann and Peter Donovan.

attendant risks. But now that age has passed and a new age is set to commence. The technology is firmly in hand and fully developed. It is at last time for the full human colonization of the seas to begin!

The undersea colony where the first families will settle is an evolved unit of human settlement that first begins with the Base and then the Outpost.

UNDERSEA BASES AND OUTPOSTS

"Tranquility Base, here. The Eagle has landed."
Apollo 11 establishes the first human base on the moon.

Undersea Base

T he first human presence in a field of exploration may typically be described as a 'base'. As demonstrated in the transmission back to earth, the first human presence on the moon was named as such. A base can be a temporary location just to reconnoiter the surrounding territory before pulling up stakes and moving on, or it can prove to be a more durable location and evolve into an outpost, as described in a moment.

A base is generally manned by explorers in a new area and functions as a center for operations as they survey the surrounding territory. Such was certainly true with the Apollo lunar bases established in six locations around the moon's surface. At the end of each exploration sequence, the astronauts left these areas and returned to earth with no stated intention of ever frequenting the abandoned base of operations again as the United States explored the surface of the moon. These bases were simply used as a point of discovery, and then effectively abandoned.

In mountaineering, a temporary encampment established as the team moves up the mountain is also referred to as a base. Again, they are transitory locations where supplies and people are staged as the team makes its way up to the summit. At the end of the exploration sequence, they are abandoned.

In manned undersea habitation this pattern has been the case in many exploration endeavors, particularly the early operations. After the missions ended in the established bases, the habitats were pulled up to the surface and the bottom cleared as though there had never been any activity there at all. In the case of *Hydrolab*, it occupied bottom locations in the Bahamas and Caribbean from 1972 to 1985 before finally being pulled back to the surface. However, the *Jules' Undersea Lodge* and *MarineLab* in Key Largo have occupied the same seafloor locations for more than 20 years.

In the future, as undersea habitation teams begin to scout out locations for permanent colonies, they will proceed just as the Apollo missions did. After the candidate locations are investigated from the surface and inspected by dive teams where possible, specially designed undersea craft (such as our LNW team's submarine, the *Dan Scott Taylor II*) will descend to the location and establish a base, sending out aquanauts for a period of days to investigate the area in great detail, record their observations, and accumulate photographs for later detailed analysis. Before returning to the surface, they will also leave behind data probes to continuously collect information on the candidate locations.

When undersea colonies are established, scientific and exploratory teams will extend the colony's operations away from the main footprint of the settlement on investigations that will require the teams be absent for a period of a day or many days. These teams will travel about in their submarines, establish bases, collect data and report back to the colony's Command Center. Upon completion of the objectives, they will return home.

An exception to this definition is a military base. Many of them have evolved into small communities complete with residential areas and many families. As in every attempt to define the future, the use of current terms is sometimes fuzzy.

If we continue on with this model, it is possible that a number of bases will evolve into some less transitory presence. These evolutionary locations are sometimes referred to as an outpost.

Undersea Outpost

For this discussion, an undersea outpost is different than a 'base' in that it is an established presence that is returned to repeatedly – not typically abandoned at the end of each use and then totally reestablished. An outpost may (or may not) be 'manned' part time, but its infrastructure remains behind for the next users. An outpost for the purposes of this discussion is the next evolutionary phase in colony building.

An outpost fundamentally functions the same as a base, except its presence is established for repeated use. This requires an established system of shelter and other requirements that help to sustain the repeated visits of passing or resident teams. A good example is the scientific outpost McMurdo Station in Antarctica. Here we find a truly impressive, permanent infrastructure that includes multiple scientific laboratories, an air strip, power station, hydroponics garden, hospital, cafeteria and so forth. Multiple teams pour in and out of McMurdo Station each year in rotating staff.

McMurdo Station is an illustration of a highly developed outpost that is somewhat indistinguishable from a colony. In fact, it is often referred to as a 'scientific colony', but for our purposes, that would be in error. The difference between a colony and even the most impressive and advanced outpost is that a colony encourages a family presence with all its necessities and demands, while an outpost is reserved for the long term maintenance of the working explorer, scientist and adventurer.

Undersea Colonies

"Adults may tend to say, '...that's just not possible',
but in a child's imagination, anything's possible.
And here I was, proving it to be true."
Aquanaut Lloyd Godson after his successful 12 day underwater
mission in Australia.

A colony may evolve from a base or an outpost, but despite the historic antecedents, in the 21st century such an evolutionary progression is not necessary. In this age, we have a wealth of understanding about development of human communities and a vast reserve of knowledge about what has happened in the historic past. There is no reason why the first undersea colonies cannot be planned, developed and launched specifically as colonies based on all that we have learned before. Indeed, if the undersea colony is not carefully planned as such from the very first phase of habitat component design, then there could be much wasted effort later and it will certainly end up a poor design, by definition.

When the early fathers of our nation had the vision that Washington D.C. was going to become the capital city of one of the planet's most powerful nations, they knew that a haphazard approach to city planning would never do. So they called in one of the brightest architects in the world, French born American artist, architect and engineer Pierre (Peter) L'Enfant, to lay out the city in advance of its expansion. The outcome was the Washington D.C. we know today – a result of intelligent forethought and vision.

Such vision is, of course, just as possible in the 21st century as it was in the 19th. And we must apply even a greater degree of forethought into undersea city planning than we have on any other human presence. As was stated in the previous chapter, an undersea colony is one whose principal purpose is to establish its overall functions and purposes broadly enough to include full family units. This uniqueness is precisely what makes it different than a base or an outpost.

The reason for such expansion of purpose is clear enough: integrating the family unit into the colony ensures its longevity. The family unit is the smallest and yet most powerful unit of civilization. Any human presence derives its greatest influence and cultural power when the family unit is installed. Otherwise, the collection of humans is just a working place or a laboratory. After the family arrives, it becomes a settlement in all of its richness, diversity and authoritative societal purpose.

But before we can even address the needs of an undersea family, there are more fundamental objections to sweep out of the way. When I invoke the image of a child in an undersea environment, for some people the fine hairs begin to stand up on the back of their neck. It is one thing to take 'daring chances' with one's own life, but to pack the kids along as a hot lunch for the local Leviathan, then that is an entirely different story.

To directly counter those reservations, I purposefully selected the picture of a young boy facing his futuristic undersea colony for the cover of this book. It was the most powerful image I could find that accurately expressed the deepest meaning of this work. I purposefully represented a boy surrounded in his protective one-earth-normal environment; at home in his undersea colony. This is the precise image I wish to leave. It is not at all the supremacy of technology; not the picture of daring, courageous adaptation. Instead it represents the authority of the protected, nourished family – which is the manifest power of our future – as it fully epitomizes the potential of man's permanence undersea.

Here are three serious myths that confront us with persistent barriers that seem to endlessly struggle against the idea of man as a permanent resident of Aquatica:

1. <u>Establishing an undersea colony appears to be simply impossible</u>.

Aquatican civilizations seem as though they are an enduring science fiction tale that somehow cannot make the transition to reality. While we can openly and freely dream and envision moon bases, 1000 day Mars voyages, or even a three decade long human mission to drill beneath Eurpoa's icebound seas, undersea colonies have always somehow been beyond our ability to grasp.

Sadly, the legacy of abandonment decades before now and the long recovery from the misdirected pathways into the sea have not been helpful. In the protracted decades since the brilliant *SEALAB* and *Conshelf* expeditions, there has been no promise of ever going back and trying it yet again. Rather than seizing and building upon the lessons of the past, we have instead relegated them to a bad dream. Instead, they should be a daring and hopeful dream of building a whole new world for tomorrow's children – our children.

Building a human colony or even a new nation beneath the seas is not only possible, it is inevitable. We need to teach that wonderful fact to ourselves and not shrink back from the promise that lies before us. Humankind has always risen to its highest and best form when faced with what have seemed to be impossible odds.

2. <u>Establishing an undersea sea colony may be foolhardy and far too dangerous.</u>

It is possible that we have Hollywood to blame for this notion. Motion pictures, however inane or silly, are still teaching tools for the masses. And because of the inventive and ingenious fusion of special and visual effects and glorious sound, their ability to indoctrinate is unparalleled in human history. Most people rightly complain at this point that they are not stupid, and this is true. But unless we viewers have some experience at undersea exploration, the Hollywood magicians may be able to sneak one or two points past us all once in awhile.

For example, how many undersea plotted movies (habitats – sinking ships – submarines) actually have a good ending for every character? Almost none. Because a terrifying mass death and drowning undersea scene has always been such a box office smash, the public has grown to view any undersea scenario as the place where almost everybody invariably dies. We have all been taught that death and drowning is an inevitable and sad fact of life under the waves.

One of my favorite things to hate is the bad science that is invariably wrapped up in these films. For example, we see undersea habitats that are pressurized to depth yet, when the inevitable flood begins and people start to horribly drown, the water is spewing in through the walls and windows. This is, of course, not possible. Water can only leak in up through the open moonpool and can never spray in through the miscellaneous holes in the walls. Just like if you forced an inverted bucket underwater in your pool and then poked a hole in its surface facing bottom. Water does not shoot into your bucket. Air forces its way out of the hole. And I will not even begin to address the lack of the realities of living in high pressure environments prevalent in many of these movies.

The reason I give these specific examples is that each scene is a teaching tool. Even when they are wrong or in error, they are always excused away as '…just Hollywood.' Unfortunately, as we add these teaching tools up, bad scene after bad scene, we have all been repeatedly taught that undersea habitation, especially for the 'poor women and children,' is foolhardy and just too dangerous. Traveling through space on the end of a rocket at Mach 25 surrounded by a near perfect vacuum in a ship with paper thin walls is fine. But forget the ocean – it's just too dangerous! The teaching power of our storytelling machines is amazing!

3. <u>It is too early to begin a human colony undersea.</u>

The driving rationale behind this argument is that somehow we are not technically ready for that great step. The truth is, the first human colony could have been built and permanently manned anytime after 1950. Had we recognized the need for a one earth-normal environment, even the pioneering work of mixed gas and saturation diving perhaps would not have been necessary. Since mankind was successfully sailing beneath the earth's oceans in 1950, then an undersea colony could have been built at that time, using the exact same technologies. With the added benefits of mixed gas and saturation diving, contemporary SCUBA technologies, advanced materials, and techniques learned over decades of undersea habitat experiments, humankind is far overdue for its first permanent presence.

If we can ever get past these myths and begin thinking seriously about permanent undersea habitation, the next viable questions center on the environment we have chosen for our children to be raised in. The best example I can point to is a former exhibit of Disney's EPCOT Center that was called 'Horizons'. It was sponsored for some time by General Electric Corporation and it highlighted our world in the future. The magnificent exhibit was permeated with a central theme: "If we can dream it, we can do it!"

I thought Horizons was the best attraction in EPCOT and over the years I rode through its many displays at least 100 times. One part of the exhibit was in an undersea colony. In this exhibit, the Disney Imagineers detailed a class of school children taught by their robotic teacher. They were all suited up and ready to take a field trip outside their undersea habitat. In the scene was a visiting seal licking the cheek of a young boy as he suited up and prepared to take an outside tour of his undersea colony.

That image has hung in my mind over these years as a single, powerful representation of families living permanently in the sea. It was rendered perfectly in the Disney style – a matter-of-fact miracle of tomorrow painted right before our eyes. It blended technology with human life infused with the ultimate power of confidence: that our technology could nourish and protect our most precious assets – our children.

I realize that it was only a staged robotic scene. Yet, if you would project that scene into the real oceans, would things be different? I cannot imagine how! Nothing in the scene would change, not even the perceived risks. But if I then move on in my mind to other scenes that I have witnessed involving real children, I can tell you that there are countless environments that children are exposed to each day in every nation on earth that expose them to far more risk of danger and imminent death than an undersea colony.

As far as cultural and educational enrichments, the average classroom does not provide anywhere near the kind of opportunity to safely learn real science as depicted in the Horizons undersea colony scene.

While we were living and working undersea in 1997 and 1998, we made frequent telephone connections with many classrooms. After the mission, we loaded up our habitat on the back of a truck and lowered it into the parking lot of many of the schools we had made connections with. It was designed for that very purpose.

These kinds of priceless interactions between Aquanauts and the nation's students were coordinated by Miami-Dade County teacher and Aquanaut Mark Tohulka. Mark is an award winning, exceptionally talented educator with a passion for undersea exploration and blending it with his classroom teaching of marine sciences.

On one occasion, Claudia was touring a group of four children inside the habitat when one pre-teen girl looked to her and opened up her heart. She said with wide-eyed amazement, "I have always disliked school and science until now. But now I think science is the coolest thing ever!" It was a remark that has hung with us over the years. It was the turning on of a light in the heart of a child – the kind of light that may never be extinguished. It is that kind of power that undersea settlements have on children – the kind of influence that changes and focuses the power of their minds.

The undersea colonies of the future will be designed to integrate the full needs of families into their structure. Without these functions they cannot, by my stated definition, become 'colonies' and they remain mere outposts.   Undersea colony services will include a full array of family oriented functions such as schooling, group entertainment, medical services, a wide range of food choices though a merchant based selection mechanism and socialization venues to name a few.

An education system will be a required necessity before the first child is admitted as a full member of an undersea colony. However, meeting the educational needs of children in undersea colonies may not necessarily resemble the traditional schoolroom environments of the past century.   They will almost certainly utilize the most advanced teaching techniques such as computer based online education, pre-recorded classes by some of the world's top educators, online testing, plus local assemblies of children of a wide range of ages and classes for field trips and full socialization.   It will probably resemble a mix of advanced 21st century home schooling techniques blended with the 'open classroom' systems in use today that rely heavily on media and not necessarily one teacher per age segregated classroom interfaces.

Medical services for a colony with families will require either the permanent onsite resident services of a family physician or an M.D. that maintains a regular rotational visit to the colony at least once or twice per week.   In the future, as colonies become established and develop into cities, full on site medical facilities will be common.

When we consider the idea of available food services, the discussion immediately broadens into one of merchant services within an undersea colony.   As we examine the concept of an undersea colony, it is essential to understand that the colony will at first represent – as well as resemble – a surface settlement in miniature.   In any surface community there are a wide range of choices available in the context of a merchant based economy. For example, as I examine my choices for food services, I can not only select from a vast array of foods, I can also select from among a wide variety of possible merchants. I have this immense decision making power because of the combined influence of our economy and societal freedoms for every aspect of our lives. Such freedom opens up a variety that is only limited by the strength of the economy.

In an undersea colony, such wide choices are automatically limited by space and the merchant's economic choice of locating in the colony. They will establish outlets there based almost solely on cost effectiveness. Obviously, the smaller the colony, the lower is the incentive for the merchant. Therefore, the colony could possibly implement its own merchant exchange system to allow a greater freedom of choice within its established and limited infrastructure. The idea is to allow there to be a freedom of choice and that freedom is enabled by a merchant economy, even if it is layered artificially from the shore based economy.

More concisely stated, a 'store' is established by the colony by purchasing goods ashore and then reselling them to the colonists; presumably for just enough profit to keep the process going and not a cent more. In many historic venues, this very method has been called the 'company store'– a much abused system that used customer employees that it intentionally indentured as a captive corporate income stream. Eventually, as the number of colonists increases, the need for the company store is naturally replaced by individual merchants and the development of a free economy. It is highly recommended for future colonists that if the 'company store' makes more than 0.5% profit, that the store manager be flushed out the nearest airlock.

As we have reviewed all these ideas, the thought naturally occurs to many people, and rightly so, "Is there even a market for this discussion? Is this book just wasting good paper on a meaningless pipe dream? Is there actually any sane person who would truly move his family into an ocean colony as a permanent resident? Are there any volunteers who want to be first?"

Obviously, Claudia and I would certainly volunteer – but our kids are all grown and gone now and it is unlikely any of them are going to loan out the grandkids permanently for more than a weekend visit or a week in the summer. So are there any nuclear families at all that would actually raise their hands and be the first to go?

In 2007, the largest family ever to become certified as aquanauts together was registered as guests at the *Jules' Undersea Lodge* in Key Largo. They were the Canadian family of Tim Novak, his wife Shelley and their three sons: 18 year old Brian, 16 year old

THE NOVAK FAMILY OF AQUANAUTS

Eric and 14 year old Ian. I asked Tim what he felt were the crucial factors to be contemplated by families considering permanent residency in Aquatica. This was his response:

"To consider permanently residing in an undersea colony we would need to discuss the employment roles of the adults, the educational and recreational facilities for each family member, the quality of life in regards to privacy, comforts, and reasonable access to medical care and external commodities.

Both adults, in our case, would want to be employed in a useful aspect to the colony in a capacity fitting their careers. The parents should feel that the opportunity enhances their career experience and that remuneration is reasonably equivalent to topside employment. Hours of work would have to be reasonable in order to allow sufficient personal and family time equivalent to similar topside situations. Activities for the adults and children, in addition to education, would include access to age appropriate entertainment and physical recreation.

Sufficient room in the domicile would be necessary to allow the family to enjoy life and pursue hobbies requiring at least a modest amount of shelf space without feeling cramped (model making or small aquaria for examples). As well, family access to the maintenance shops and marine labs for personal projects would also be appreciated.

Mail order access to certain products and consumables in addition to a well stocked general store would also be necessary.

As an analogy, all the above is similar to how company mining or forestry towns operate in remote areas. Such communities typically allow more comforts that attract families than would a scientific outpost similar to those in Antarctica."

Tim's real-world answer clearly demonstrates the difference between an outpost and a colony. Here Tim's concerns can be summarized as 'room and enrichment'. In other words, the colony must be designed and oriented toward meeting the needs of family existence as well as its other purposes.

I then asked Tim, "What would be the best thing about such a life with your wife and children?"

"The best thing about such a life would be the shared sense of adventure, community and purpose and the opportunity to raise one's family in a safe and stimulating environment."

Again, his answer centers around his responsibility as a parent and husband to provide a safe and enriching life for his family. But he also views the undersea environment as one that could meet those needs.

Just to be fair and balance the discussion, I then pushed Tim to the 'what if' limit with this question: "What would be your greatest fears about such a move?"

"As the individual needs of each family member evolve, a point may be reached where not everyone's needs can be met within the colony. This would result in the necessity for one or all to relocate topside. Of course, that is no different than most topside communities."

It was interesting in his answer that Tim felt no need to articulate his 'permanence' in Aquatica as an obsessive need for total separation from the land. Tim expressed a sense of equal connectedness and total freedom of movement anywhere on earth even though his permanent address would be Aquatica. His stated priority was that the needs of his unified family come first in deciding where they will live. What I find most interesting in his answer is that it comes from a balanced, healthy, stable head of a real family and yet, there seems to be no sense of hesitance in holding the rational discussion and consideration of moving his family to a permanent undersea address.

I then asked him, "Since you most probably hold the world's record for the largest family group of aquanauts to live undersea for 24 continuous hours, do you even for one moment also fantasize about taking your family down and living permanently in the oceans?"

"Yes. There have been a few science fiction films made about families living together in an undersea or outer space based home. The life portrayed, albeit fictional, was very intriguing and attractive."

When I felt I understood Tim, I asked him, "What do you think your family would say if you really suggested that?"

"I have suggested just that, and the boys gave an immediate positive response, while Shelley said, 'Sure, as long as systems to prevent accidental flooding are sufficient.'"

I was somewhat impressed that Tim actually popped the question to his wife in a real world, sincere way and she had a response I could actually quote. But not being entirely satisfied by the adult responses, I asked Tim if I could talk to the youngest of the boys, 14 year old Ian.

DC: "Ian, what would you think about moving permanently into an undersea colony?"

IN: "It would be a very cool, interesting, and fun experience."

DC: "What would be the most interesting thing about it?"

IN: "That it's like a base underwater. You could go SCUBA diving for a long time and come back without having to decompress."

DC: "What would your friends think about it if you did?"

IN: "They would think it was interesting and that I was pretty lucky to be able to experience something like that."

The whole point of this discussion is to demonstrate several essential ideas:

1. Rational, educated, stable, normal, balanced people have serious thoughts about permanent residences with their families undersea.

2. An undersea colony is not just a base or an outpost. To be viable, it must take the families full range of enrichment into account before it is ready for families.

3. Undersea colonization does not mean separation from the land. Individuals and families retain full freedom of movement, which is ensured by the principle of a one earth-normal environment.

But soon, just like on land, colonies eventually grow up and into fully functioning cities.

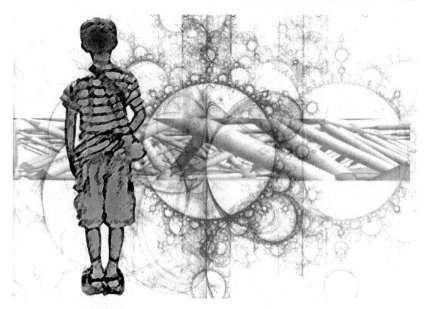

UNDERSEA CITY

"You will never be able to discover new oceans unless you have the courage to lose sight of the shore."
Hannah Whitall Smith

E | ventually the undersea base will grow into an undersea outpost. The outpost will become a colony and the colony will become a full undersea city. That city differs from a colony in its range of services and the robustness of its infrastructure. The formal definition states, "A city consists of residential, industrial and business areas together with administrative functions."[a] It can get fuzzier than that, but there is no need for adding complexity. That definition will do just fine. An undersea city will consist of residential, industrial and business areas together with administrative functions – just as they do on shore.

If you can envision a colony as a group of undersea structures that are connected to fulfill all the functions that allow families to live there permanently, then imagine a city as a large colony whose operations are segregated into functional areas. I think you can see why there is a requirement for city planning even before the first habitat is placed undersea. Eventually there will be a requirement for segregating the family living units in communities separate from the industrial and commerce sections of the undersea city.

I once visited a small country and checked into a downtown hotel. As I walked to the hotel, I had to step off the sidewalk and onto the street on several occasions to avoid factory and industrial services activity along the street that had slid their work out over onto the sidewalk. In other words, unless I wanted to be sprayed by hot welding and grinding fragments, I had to get out of the way of their operations. Upon arrival at my hotel, I rode up a scary elevator to the 10th floor – sans fire escape. As I looked outside my hotel window, I could see a residential neighborhood just below intermingled with the industrial and commerce facilities.

It made me realize for the first time in my life how essential proper city planning really is. Without proper city planning, the city is forced to compromise like the one I visited so that there is no order, harmony of function, or even safety. The only other choice is to disassemble the growing colony and reassemble it as the city branches into its functional divisions. That is almost impossible and is certainly not economically viable.

The Undersea City's Residential Areas: In an undersea city, there will be areas that house families. They will be segregated because families form communities and communities thrive as a connected social structure. These undersea residential areas will consist of individual family living units as well as small shops, community centers, schools and churches, all linked by transportation hubs.

The Undersea City's Industrial Area: The undersea industrial complex will be segregated for reasons of safety and function. These areas will include factories, laboratories, research centers and logistic receiving locales. The industrial area will contain community oxygen producers, community CO_2 scrubbers and other city life support equipment. They may also include arrival

JACQUES ROUGERIE'S WONDERFUL VISION OF AN
UNDERSEA CITY BRANCHING OUT FROM ITS PRIMARY HUB.
WWW.ROUGERIE.COM

and departure points for submarines, logistic pods and colony-to-surface elevator systems.

<u>The Undersea City's Commerce Area</u>:   This area represents the 'undersea mall' of the city. Perhaps no better model exists in our culture of this area today than the common community shopping mall. It is an economically viable way, especially in an undersea setting, of accumulating all the commerce of the community in a single, easy to reach and convenient venue.

The first logical question is, "When does the undersea colony turn into the undersea city?" No one knows the answer to that question and it is perhaps irrelevant. What we are witnessing in these pages is the visible evolutionary process of the human presence developing undersea in terms we can all understand, based on our knowledge and history of the development of land based colonies and cities.

As we look even further ahead, we know from past history that human expansion does not stop with cities. Eventually there are many cities and these cities can join together to form a sovereign entity called a nation.

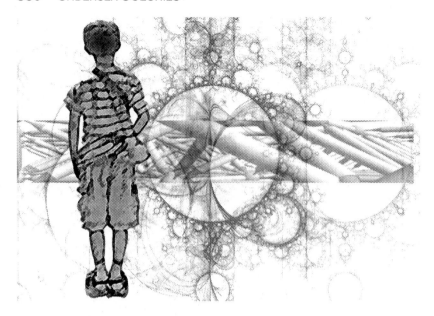

UNDERSEA EMPIRE

*"The use of the sea and air is common to all; neither can a title to
the ocean belong to any people or private persons, forasmuch as
neither nature nor public use and custom
permit any possession thereof."*
Queen Elizabeth I - 1533

U ndersea nations cannot, by international legal definitions be staged near shore. Most established nations claimed the waters adjacent to their shores as their sovereign territory, at least to the three mile limit. But in the 20th century, this began to change. By 1967 only 25 nations still used the old three nautical miles limit; 66 nations had set a 12 nautical miles territorial limit; and eight had set a 200 nautical miles limit.[a]

The 200 mile limits were labeled 'Exclusive Economic Zones' by the United Nations Third United Nations Conference on the Law of the Sea in 1973. They never anticipated undersea nations at the time so the international law grants the adjoining nations the right to exploit all the natural resources of those waters.

However, any other nation is allowed full freedom of navigation and over-flight, subject to the regulation of the coastal states in as close as the 12 mile limit. Navigation implies, 'keep moving', and does not mention the right of undersea squatters to stay and build under the surface.

Closer to the shores, the 12 mile limit is defined as 'Territorial Waters'. Inside this limit the powers of the state increase, and it is free to set laws, regulate any use of the water and use any resources in the water. Further, there are more restrictions in passage through these waters. Vessels, surface and submarine, are allowed 'Innocent Passage' which is defined as passing through waters in an expeditious and continuous manner which is not 'prejudicial to the peace, good order or the security' of the coastal state. Here you can safely assume it means no military operations are allowed inside these waters, including spy ships and foreign military submarines, without invitation. Again, no mention was made of undersea settlements.

From my reading of these international laws, I do not think any nation on earth would allow the sovereign declaration of an undersea state within 200 miles of its coastline. That, by geographical constraint would limit any undersea nation to being tethered to the bottom since depths beyond 200 miles may not be practical for a seafloor site. And yet, that limitation by no means crowds the field. It still leaves more than half the globe wide open to undersea nation building.

Why would any collection of undersea cities even be interested in sovereignty? The answer to that question is as old as human civilization itself. If I may be permitted to answer a question with a question, why has *any* nation wanted sovereignty? The reasons are wide and varied. The United States was founded for reasons of religious freedoms and expanded freedoms of governance, speech and economic development. But every nation on earth has its own individual set of motives for why they desired to become a sovereign, independent nation.

Nations have never appeared overnight. They have typically been formed after many generations of colonial rule when the governance of the sponsoring state becomes a burden and its demands (tax) exceed the reasonable expectation of the residents of the colonies. This is sometimes, but not always, followed by harsh disagreements and occasionally war. But distant sponsoring states can rarely hold onto remote colonies and eventually they acquiesce – either at the end of a bloody disagreement or through simple economic pressures – and a new, sovereign nation state is born.

For what purpose will Aquaticans wish to establish a new nation state? One reason will be to protect their resources (and thus their economic and military control) from being plundered by the land based nations. The undersea regions of the earth are at least as resource rich as the land areas and there are some indications they may be more so. That means that roughly three quarters of the viable resource richness of the earth lies still undiscovered in Aquatica.

Since the Aquatican civilization will naturally (by definition) claim the undersea world as its own technologic base, then recovering those resources will be far easier for their society than for a land based nation. I would like to extend this argument one step further. Even though it is difficult to perceive this now in an area whose population is exactly zero, it will therefore be the Aquatican civilization that will become the most resource rich and hence the most powerful nation or nations in the future of this planet.

It is unseen now, lying just ahead of us in time; therefore it is difficult to relate to the concept. Yet, there is no argument in history – the richest nations become by definition and process also the most powerful. Aquatica holds the wealth of the world in its dark and unseen three dimensional seascapes. Opportunity for dominance now awaits only the outcome of the endgame.

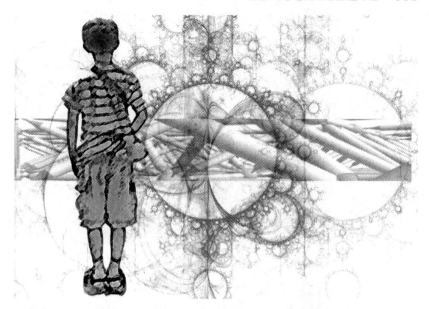

ENDGAME

"Instead of gazing down through water buckets and glass-bottomed boats, in addition to watching the fish milling about in aquariums, get a helmet and make all the shallows of the world your own. Start an exploration which has no superior in jungle or mountain; insure your present life and future memories from any possibility of ennui or boredom, and provide yourself with tales of sights and adventures which no listener will believe - until he too has gone and seen, and in turn has become an active member of the Society of Wonderers under-sea."

William Beebe.

In this book we have traveled all the way from the present, when the population of Aquatica is zero, to the building of undersea empires in the distant future. While many will not agree with this assessment – and most have not given it any thought at all – it is, in fact, inevitable. There are no places where humans have not settled from the coldest Antarctic and Arctic regions to the hottest deserts on earth. Today we are planning and in the initial stages of building spaceships that will travel to the moon and Mars. Their specific purpose is to explore these remote and distant places with the intent to scout the locations where humans will establish permanent bases, outposts and ultimately colonies. The identical process is now ongoing through at least one private enterprise for undersea colonization.

The question that each nation must ask itself is this: Are we willing to participate in this expansion? Yet, the question runs far deeper than mere participation. With the expansion that is represented by each human colony is also the extension of that nations' ethos that includes all of its beliefs, its cultural history and all the rich diversity that makes any people what they are. Historically, nations that have shrunk back from expanding have all been swept away in the wake of those who do. They lose not only their sovereignty; the greater loss is that their cultural identify is ultimately erased by more energetic and purposeful nations with the impulse to expand. Exploration and discovery is never a choice of whether to participate or not; it is a decision to either consciously terminate the culture or carry it forward. Exploration may well be born in a spirit of geopolitical competition, but it is physically expressed in tomorrow's real world articulation of cultural power.

The question represented by this book looms larger and more immediate to countless people of all ages. Many eyes will scan these words and each person attached to them will be faced with a choice: Do I move into this new frontier or do I remain fixed to the apparent safety of the land?

The choice will never be easy. I know that Claudia and I have agonized over its reality in our lives and what it will ultimately mean for us. We have come to full terms with our destiny, wherever our Maker decides that will be. We are ready to step off *terra firma* into the blue dominion that lies just before our feet and never return. It is true that we will retain our freedom of movement and our capacity to travel to wherever on earth we so choose, but we are equally ready to permanently change our address to a place that does not exist today, but one that we are convinced soon will!

We have had the advantage of having lived there before with our family in tow. We know what it is like and we are fully aware of the risks as well as the benefits. It is not for everyone, but neither was North America in the 15th through 17th centuries. Adventurers, explorers and colonists have always been a peculiar lot, willing to exchange some comforts for the raw adrenaline of living in a new frontier. There is ultimate freedom to be had there. But, just as in any other human enterprise, there will be risks. Yet with great risk comes great opportunity, and with great opportunity comes the possibility of great reward for both nation and individual.

The time has finally come. After more than five millennia of human civilization and a hundred billion human lives, the first people are poised and ready to move into a place where no one has gone to live out their lives before. They stand now, at the seashore; the white foam of the single, great planetary ocean laps gently around their bare feet, the green water of the sea folding before them in wave after wave as it always has. Never before have humans stood with their eyes cast at the global sea and its vast undersea dominion and viewed it as a place to be called home.

Until now...

As you read these words, the plans are being laid, the next generation of undersea habitats designed and built, the new transportation systems are under construction. Many future Aquaticans know who they are and they are very busy. The first day of a new empire lies just ahead. It is no longer a matter of decades; it is now but a matter of days.

THE CERULEAN DOMINION

One day in the Future...

Cerulean Dominion
Location: 345 miles WSW Miami, Florida
An Undersea Colony anchored at 120 feet beneath
the surface of the Atlantic

A aron Seven sat nervously in his seat as the small helicopter approached the tiny dot floating on the surface of the North Atlantic ocean, more than five thousand feet below the aircraft. The pilot had notified Seven, the sole passenger, that they were making their approach to the S3 Platform that served as the staging and interface docking point for the undersea colony, *Cerulean Dominion*, also known as *Dominion*, below. Even though Seven knew that he could never see the colony located some 120 feet below the ocean's surface, he strained in vain to try to catch a glimpse of it anyway.

The helicopter dove down sharply toward the platform as Seven's stomach reacted to the faster-than-any-elevator feeling of losing altitude. As he stared out the bubble window, he could see the platform growing in size from a dot to a postage stamp and eventually to a more comfortably sized area with a landing pad on one end.

The flashing strobe in the center of the landing pad blinked repeatedly as the helicopter began to slow its descent to settle onto the now appreciably sized platform noticeably heaving in the ocean swells. Seven gripped the stuffed bunny in his hands as they approached – a gift from the surface for his daughter, Rachel.

Eventually, the full view of the S3 Platform emerged as the pilot lined up the helicopter to land. It was the colony's only connection with the world above – a submergible platform that could be raised and lowered as they pleased and one that served many invaluable functions for the 432 humans recorded as permanent citizens of the largest undersea colony on earth some 20 fathoms below.

The chopper finally touched down so gently that Seven whistled and said, "Wow! Drake! That was the smoothest landing ever – and in a meter and a half swell! I'm very impressed!"

"Thanks, Dr. Seven," he responded. "I appreciate that!" Seven was his boss, as well as everyone else's, being *Dominion*'s founder and director.

Seven stepped out of the helicopter and immediately felt the sea's gentle raising and lowering of the platform below him. He looked out over the ocean to the horizon and with expert eyes estimated the swell was not a long period wave and would not be felt in the colony below.

"Dr. Seven, welcome home!" said a small-framed man sharply dressed in starched coveralls wearing mirrored aviator's glasses as he approached the helicopter whose blades were still winding down.

"Thanks, Nick," Seven responded and then turned to the pilot. "Hope to see you on the next trip, Drake. Have a safe flight back."

"Not today," Drake responded with a sly grin. "Got a hot date – down there!" he said pointing toward the colony. Then he eyed the stuffed bunny Seven clutched in his fingers. "That for Serea or for Rachel?"

Seven laughed. "It's for me."

Drake opened his mouth to retort, then stood there considering for a moment, turning red.

"Okay then, twist my arm. It's for Rachel," Seven chuckled bailing the young pilot out of his dilemma.

He looked relieved, then inquired, "May I catch a ride down?"

"Sure, Drake. Who's the lucky woman?"

"Who said it was a woman?" he responded instantly with no smile.

Now it was Seven's turn to blush, speechless.

"Okay then, twist my arm, sir. It's Jan Mason!"

"Touché!" Seven replied with a chuckle, the face of Jan Mason, the brilliant microbiologist, springing instantly to mind.

"Your boat is ready, Dr. Seven," Nick broke in, handing Seven a palm sized computer. Seven looked at his screen and all the data about his four man submarine, the *DST* VI, and its systems were already flashing on it before him.

"Powered and ready, I see," Seven remarked, scanning the numbers.

"Yes, sir!"

"Good job, Nick."

"Thank-you sir," he responded.

But Seven noticed the twitch in his right hand as he forcefully held back a salute. Seven looked at Nick and smiled. "Good reflexes. No salutes here, remember. You're certainly not in the military anymore. Oh, and log in Mr. Drake Logan, please. He'll be my guest tonight."

"Already done, sir," Nick responded too quickly.

Seven looked at Nick then at Drake. "Do I sense a conspiracy?"

"Ahhhh..." Nick began nervously.

"It's my doing, sir," Drake blushed. "I just asked his opinion last week and he said you probably wouldn't mind..."

Seven laughed loudly. "Tell me, my young friend, which motivation makes you *Dominion*'s most frequent visitor – your love for Aquatica or your hormones?"

"Both, sir," Drake responded without hesitation, nodding his head with certainty. "I'd say definitely both."

Seven smiled. "Get in the boat and let's head down before I have you and your fellow conspirator keel hauled. I can do that, you know."

Drake smiled at Nick and flashed him a thumbs up.

Nick shook his head slowly. With a sigh he said to Seven, "Sorry, sir."

Seven smiled at both of them. "You guys are good – very good. I approve. However, next time let me in on the plan a bit earlier than, 'Here I am!'"

Seven led the way to the opposite end of the platform and down one level to a compartment at the water line. He opened its hatch and there facing him was the familiar sight of the *DST* VI's conning tower slowly bobbing up and down in the water. Seven stepped across to its platform, up two steps and swung open its heavy steel hatch with a twist of its upper handle. With a wide swing of his right leg, he stepped over to its interior ladder and climbed inside. Behind him Drake followed and expertly pulled the hatch closed behind him with a clang.

Seven slid into his seat and clipped the handheld computer into its cradle. It was not only a part of the colony's paperless society, it also served as his navigation unit that could guide the *DST* VI down to a hands-off safe docking, if required. But Seven was an accomplished submersible pilot and always preferred the manual approach. His fingers danced over the toggle switches as he checked his lists and prepared the sub for departure.

As Drake strapped himself in, Seven opened the communications circuit. "Nick, release us. We're ready to dive!"

"Awesome!" Drake reflexively whispered, his eyes dancing over the multicolored panels before him.

"Umbilicus release," came the voice of Nick, followed by a resounding clang as the mechanism let go.

Immediately, Seven reversed the engines and the 24 foot submarine backed safely away from the S3 platform rising and falling in the ocean swells. Simultaneously, Seven cycled the ballast valves and frothing water immediately filled the view of each porthole.

"Styling," Seven snapped into his lip mounted microphone, indicating he had commanded the sub to turn in a tight circle as it headed down. This enabled him to immediately control the sub's undersea attitude and depth without having to wait for a linear motion which would have taken him away from his target.

The water turned immediately blue as the North Atlantic reflected into their windows on this brilliant, clear, sunlit day. The beautiful, deep blue was why they had named this colony *Cerulean Dominion* in the first place. Seven and Drake looked out the sub's forward windows and each caught a glimpse of the colony's far eastern edge as the sub rotated downward.

"That is just so unbelievably cool!" Drake said in a low voice, looking over to Seven. "I have to ask you a special favor, Dr. Seven," he added abruptly.

"The answer is no, Drake. We've had this discussion before," Seven responded flatly. "There are no open positions in *Dominion* and there is a line up of at least a thousand families wanting to get in. I'm sorry. As much as I like you personally, I just can't make an exception to our rules."

Drake did not pause. "But what about this?" he asked, fishing a diamond ring out from his pocket. "Would this make a difference?"

"I'm already married, Drake; you know that," Seven responded with no trace of a smile.

"Very funny, Dr. Seven. If I give this to Jan tonight, and she accepts, would that make any difference?"

Seven stared back at him, speechless. By his own rules, spouses of colonists were automatically admitted to the colony.

"Listen, dear boy. I happen to be very good friends with your fiancée. And if I find out that your intentions are less than honorable...."

"What do you mean, Dr. Seven?" Drake asked with a deep frown.

"How do I know that you're not marrying her just so that you can live here?"

"Please..." Drake protested. "If she dumps me, then I have to move out. Isn't that incentive enough to become the world's best husband?"

"Wrong. If she dumps you, I'll personally flush you out the nearest airlock," Seven responded with an unreadable expression bordering on severe. Then he abruptly smiled, "Good luck. If she says yes, you're in – after the wedding."

"Well, you can marry us. Am I right?" Drake asked.

Seven stared back at him and blinked. "Yes I can. Are you sure about this?"

"Yes. As sure as I have ever been about anything."

"And Jan?"

"We'll have to wait and see if she says yes. Will you do it tonight so I don't have to go back?" Drake asked.

"Doesn't the bride have some say in this? And what about the helicopter? Who's going to fly that back to the mainland?"

"I can keep my flying job and live here!"

Seven sighed. "Drake, you drive a hard bargain...but, it sounds good to me!"

Drake smiled and popped the ring back into his pocket.

"You had this all planned out, didn't you. You and Nick..." Seven said shaking his head slowly.

"And Jan..." Drake added with a grin.

"Oh, and now *she* is also implicated in this conspiracy," Seven sighed. "Let's take the long way back. I want to show you your new home."

"Dominion Command Center," Seven said into his mike. "This is the CD in DSV six. We're going to take a swing around the perimeter and approach the docking bay from 270."

"Roger that, Dr. Seven. Welcome home. We'll notify your wife of the delay. She and your daughter are waiting at the dock."

"Thanks CC. We'll not delay more than five to ten."

"I want you to see the colony from a close fly-by perspective," Seven said to Drake, banking the sub away and out into the blue void.

DST VI headed away from the colony. Seven could see nothing but an azure glow ahead of him, suspended in the open sea. 120 feet above him the surface interface was unseen from this depth, and below him, the blue dropped off into the blackness of the abyssal depths. The colony was suspended by anchors set at great depths that could be adjusted to position it from the surface to any deep depth they pleased.

Seven adeptly turned the boat and soon the colony came into view from a distance. He steered the craft so that they approached it from its edge. The sight was startlingly beautiful. The edges of the massive platform onto which the habitats were attached blinked in multicolored lights and each mounted structure's windows glowed with interior illumination.

The view of the colony grew in their windows as they approached; its functional layering became apparent. It was constructed in three horizontal tiers and designed so that as the colony grew, layer could be added to layer without having to expand the width of its base - although even that was possible in the future. It was strikingly beautiful and awesomely functional, designed by Seven and his wife Serea.

It was devised with a specific purpose and that purpose was initial, immediate functionality, and ultimate expandability. *Cerulean Dominion* was the most carefully designed and planned human settlement in history. Its striking lines, form and beauty were also intentionally incorporated.

Seven steered their craft below the structure and headed toward its edge. "I want you to see our shock absorbers," he said. The sub descended to a hundred feet below the colony's base at one corner and before them loomed a massive cylindrical structure into which the anchor lines from the deep were intertwined.

"We have four of these absorbers that allow us to ride out long period swells and surface storms. They're designed to permit the colony to ride through the energy as it passes and not endure the impulse and inertia of the dynamic energies during each passing swell or surge."

Seven passed to within 30 feet of the huge absorber, and then slowed the craft so Drake could see it in detail.

"That's an example of what we call an 'enabling technology' that allows us to suspend ourselves from the seafloor. Without that technology, these kinds of undersea structures wouldn't be possible. Another enabling technology is the ability to live in structures that are pressurized to one atmosphere. Without that and the various pass-through systems, we couldn't make this work at all. It would be impractical to have a human presence here and far too dangerous. But with the enabling technologies, *Dominion* is safer than any other human settlement on earth."

Seven then circled the craft out and up. He passed within 50 feet of the middle occupied section of the colony which, by policy, was as close as he was allowed to bring his vessel.

Drake stared out the porthole in open mouthed wonder. Seven, too, never tired of this view. It was as magnificent a sight as there was on earth. Here, suspended in the deep blue void, was a creation as spectacular as any other ever fashioned by the hand of man. It was truly a wonder of the 21st century world.

"Better get to the docking hub before they send out an escort," Seven quipped, banking the craft away from the structure and styling down to its deepest module, the colony's dock. As they approached, Seven could see five other submersibles docked side by side, their conning towers poked into their individual receiving bays.

"DST six on the approach," Seven announced as his fingers gently controlled the smooth transition forward. As he drew closer, he could finally make out Serea and his young daughter, Rachel, peering out at him and waving through a very large domed viewport. He flashed the conning tower lights at them.

Seven then focused his attention on the display before him, following the sonar guided entry path to the dock. Eventually, as the display's brilliant green crosses merged and became one, he snapped back on the submersible's engine switch. "Engine stop. Thrusters up 15 percent."

The sub gently slid into its rubber encased dock. He could hear the hiss of the air being equalized in the docking chamber to one atmosphere.

"Hard dock, Dr. Seven. Pressure nominal. You have permission to debark. Welcome home!"

Seven stood immediately and turned to Drake. "Welcome to your new home, lad," he said extending his hand.

Drake took it readily and smiled exuberantly, pumping Seven's arm. "Thanks, sir," he said. Then, as though he could not contain himself, he embraced Seven tightly.

Seven allowed the emotional moment to pass, then politely pushed him off. "You're the perfect colonist, Drake."

"Huh?" Drake answered.

"You're exuberant and excited. Those qualities are actually more valuable here than an advanced degree!" Seven then snatched up the stuffed bunny, bounded up the ladder and popped the hatch open. There before him stood Serea, little Rachel and none other than Jan.

Seven stepped over, knelt and handed the bunny to Rachel who received it and hugged him. "Thank-you, Daddy! I love you!"

"I love you too, sweetie," he said, planting a kiss on her cheek. He then stood and gave Serea her gift in the form of a deep kiss.

"Welcome home, dear," she said with a smile.

"Serea, I want to introduce you to..." he began, lifting his hand to where he thought Drake would have been standing. But in the distance, on the pier, he could see Drake and Jan entangled in a more than passionate embrace.

"Well, I guess this whole thing was indeed a sinister plot, after all," he sighed.

"Whatever are you talking about?" Serea asked with a wry smile.

"You, too?" he asked. "You were in on this?"

She just shrugged.

"Figures," he said with exasperation. "Well," he added with a sigh, snatching his daughter off the deck and into his arms, "we'd better get onto the wedding. We can have it in my office in an hour."

"Wait a minute! We have to plan these kind of things," Serea protested.

Seven sighed and closed his eyes, shaking his head slowly. "Whatever."

It was one thing to plan and build this magnificent undersea colony that was rapidly becoming a city. It was another thing altogether to figure out such far deeper mysteries as love, women, men and their families and how they all fit together...

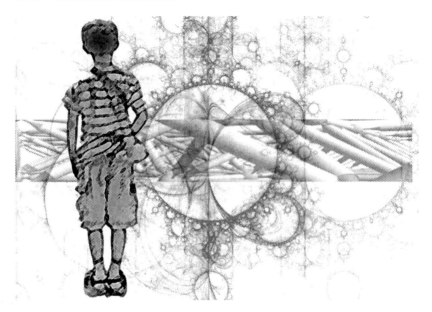

A NOTE TO THE PROFESSIONAL DIVING COMMUNITY

*"If my critics saw me walking over the Thames they would say
it was because I couldn't swim."*
Margaret Thatcher

B ecause I have so many personal, close friends in the
professional diving community and because they built
the system in which I and others have benefited to
date, I wanted to be certain that this work does not come across as
a critique of their essential work in the past or what lies ahead in
the future. The professional diving community brought us to
where we are now and they will be responsible for building the
colonies in the future. So, allow me to emphasize that this book is
in no way meant to criticize the current professional and mixed gas
diving community.

In the past, nearly all undersea habitat activity was
accomplished by these professionals to the point that the two

entities were nearly indistinguishable. But this book looks forward to the time that most people who live and work undersea will not be diving professionals and will never in their lives taste mixed gasses. That does not mean that there will not be the need for them, because as the undersea communities grow, there will actually be more need for professional divers and the profession will expand greatly! Without them we could not have achieved what we have, and without them we cannot progress another day forward.

This scenario is akin to passengers on jetliners. Today, millions of people fly around the globe each year, yet almost none of them are pilots or know the first thing about flying a jetliner. But early in the history of aviation, nearly everyone who flew in an airplane was a pilot sitting in an open cockpit, directly exposed to the elements.

Today, in undersea habitation, every one who lives and works in habitats is a diver of some sort with at least some proficiency in diving and being directly exposed to the sea as they come and go. That will soon change. This book is written about an undersea world that has not yet been born – a world that will soon be populated by people who never even get wet, most of whom will never don diving gear and some who will not even be able to swim!

This book openly discusses why undersea habitation by large numbers of people has not progressed, and many of the reasons are related to the fact that few people wish to live in tiny cylinders and undergo risky and complicated decompression while talking like Donald Duck. Aunt Miriam does not even want to discuss taking her children – six year old Billie, eight year old Frank and five year old Misty – to live permanently in this kind of uncomfortable and often precarious environment.

In this book, I have taken the side of Aunt Miriam and the kids.

Yet I have not criticized the professionals in the process. That would be very foolish! Were it not for the professional divers and those who risked their lives and lost their lives in preparing our way, we could not take the next step. And because of their continuing expertise and leadership built up over decades of sacrificial, path-finding exploration, the new undersea colonies will be made possible by the creation of the new world to come, fashioned by their handiwork, expertise and love for what they do so very well.

SAFETY NOTES

I can just imagine reading this book as a lad. I do not believe that I would have even finished its final pages until I would have been busy building my own undersea habitat. I am certain that would have been my response! So please allow me a moment to speak to you lads and lasses of all ages with that plan in mind – just as I would have spoken to myself, armed with the experience that I have today.

There is one key and most important thought when approaching undersea habitats. It begins and ends with one word: safety. Safety is first, it is last and it is everything. Never, ever compromise safety. It cannot be overdone; it cannot be too well planned for; it cannot ever be allowed to take second place. No undersea habitat is ever worth even one precious human life.

An undersea habitat – whether it is on shore or under the sea – is a closed system. That means the very second you close the door, you begin depleting your oxygen and creating a cloud of carbon dioxide that you will have to re-breathe. This system is therefore a dangerous one by its most elemental definition. I would strongly suggest you take the very same approach I always have when building an undersea habitat – leave the windows and doors off until you are ready for the water.

A habitat will require 64 pounds of *in water* mass for every cubic foot of air that it displaces in the water. That creates the need for them to be very, very heavy. Whatever you use for ballast and whatever plan you have for getting it underwater will require thousands of pounds of force. These force vectors are very dangerous. Be careful.

An undersea habitat is very difficult to launch. Even the world's leading experts have had trouble getting a habitat down safely. Stay safe, whatever your approach is to this step.

Never, ever get inside a closed container underwater until your system has been checked and re-checked by an expert. Whether it is a five gallon bucket over your head or a huge metal box, do not go inside it until you have had adequate supervision.

Never attach live electrical wires to a habitat without including the advice of a licensed marine electrician. Salt water is a good conductor of electricity. One mistake can cost you your life.

Parents, if your child is building an undersea habitat, please encourage them to the utmost, yet please closely supervise them every step of the way. If you are not certain about something, stop and seek expert advice before proceeding.

Kids, let your parents know what you are up to and ask for their help! Show them this page! (Come on now – you will probably be surprised how they will respond if you are honest with them.)

Finally, remember the most important thing:

Safety is first, it is last and it is everything. Never, ever compromise safety. It cannot be overdone, it cannot be too well planned for and it cannot ever be allowed to take second place. No undersea habitat is ever worth even one precious human life.

I am very sorry to say that I am not permitted to be involved in your design. My attorney has advised me that I should definitely not walk off that plank – the ocean is way too full of hungry legal sharks who are looking for some poor author to eat alive, and I cannot afford to thereby jeopardize the future of my own undersea projects. However, by all means, approach your habitat building project with a level head, with all the advice you can glean from many sources and all your neighborhood experts. Do not allow any of them to tell you it is impossible or will require millions of dollars, either!

Be safe and do not give up!

Finally, if you are serious about building your own undersea habitat, let me strongly recommend that you obtain a copy of James Miller and Ian Koblick's *Living and Working in the Sea*, available from the Internet. Read it at least twice cover to cover before you begin. I actually had to buy a second copy – I wore my first one out! It is the most comprehensive book on undersea habitat design and building that is available and is full of lessons learned that you will not want to miss.

You can also get in touch with other future Aquaticans and habitat enthusiasts thru the Yahoo Discussion Group: Undersea Colonies. God luck and God speed!

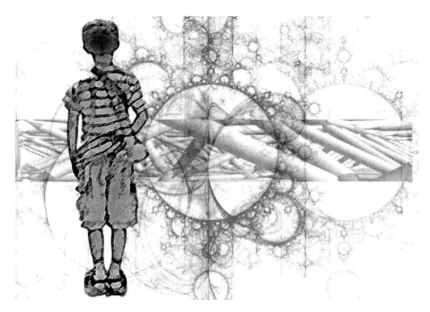

Acknowledgements

"Asking is the beginning of receiving. Make sure you don't go to the ocean with a teaspoon. At least take a bucket so the kids won't laugh at you."
Jim Rohn

T his book has been swirling about in my head for a number of years. But I can also say that my head is as at least as much a magnet as it is a generator. It consists of the hopes and dreams of many others besides myself. In the end, I hope this book is a work of enablement to actually realizing George Bond's original dream of mankind's ultimate achievement of dominion over the seas by living and working there.

If there is uniqueness here, it will be in the synthesis of other work that has gone before me. If there is originality here, it is because all originality is ultimately enabled by the sacrifice of many teachers who skillfully and sacrificially taught us how to view the world through the filters of other's accomplishments. And in so doing, we are enabled to paint a new picture empowered by dreams that have actually been generations in the making.

In the opening dedication page I gave credit to the giants who have walked before me and the giants in my own life who have made these opportunities available that led to the writing of this book, so I will not mention them again here.

My incredible wife, Charlie (Claudia) is just amazing, and continues to be remarkable, and any detective of even average talent will certainly be able to find her fingerprints on every page of this book. To my other faithful editors, Martha Smith and Susan Austin, thank-you for your quick (as usual) turn around in the last second throes of a late manuscript. To make this particular process even more spectacular, Martha Smith demanded her edit copy of these pages while lying as a patient in a hospital ward's Intensive Care Unit!

I owe a great debt of gratitude to the Yahoo Discussion Group Undersea Colonies. Their ideas permeate this book with such adventure-explorers as Australian Aquanauts Ralph Buttigieg and Lloyd Godson, and Canadian Aquanaut Tim Novak and his family: Aquanaut wife Shelley, and Aquanaut son's Brian, Eric and Ian. And I very much appreciate the suggestions offered by Alex Michael Bonnici and Shaun Waterford as well as professional diver and great encourager, Rob Bryan. I must also highlight the incredible generosity and assistance of Astronaut Duane (Doc) Graveline and the endless, non-stop support of Astronaut-Aquanaut Scott Carpenter. All of these individuals have had a hand in this book's inspiration and development.

I must also extend my thanks to the incredibly talented Spanish artist, Guillermo Sureda-Burgos whose fantastic art depicting the future of undersea transportation is found in the chapter on commuting. He has one of the most visionary and talented eyes on the planet for what the future will probably look like. And to my brilliant and very talented son, Brett, who created the models for the *Leviathan* Habitat and the redesigned DST II.

I also wish to thank the marshaled crew of the Atlantica Expeditions – some of the most experienced undersea explorers ever assembled – who are planning and scheduling the first permanent migration of humankind into the oceans of the world. Each of you inspires me individually each and every day!

And I wish to especially thank Vicki Mudd, as well as her husband, Kevin, for her more that generous expression of faith in the next great step into the oceans. Vicki opened a door to the frontier in a very spectacular way!

Finally I wish to acknowledge the rather small family of Aquanauts that have been inspirational to me just because they have joined together in this magnificent and almost unknown quest. Soon, very soon, the population of Aquatica will rise above zero. And I hope, dear reader, whoever you are, no matter how young or old you are and no matter where you are, that you, like me, will be counted in that number!

Soli Deo Gloria

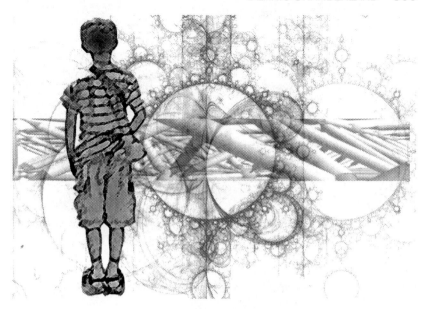

GLOSSARY

Accretion – The process of forming calcium carbonate layers over wire mesh frames underwater by passing a relatively low powered DC current through the wire.

ADS (Atmospheric Diving Suit) – See Hardsuit

Atlantica – The name selected for a permanent undersea base/ outpost/ colony by LNW Corporation off the Central Florida coast, scheduled for opening in 2012.

Advanced Life Support System – A life support system that recovers and recycles part or all of its nutrient/resources again and again.

Ambient Pressure – Considered the pressure of the water at the depth of the moonpool surface of an undersea habitat.

Ambient Pressure Habitat – A habitat that is pressurized to the depth of the water at its moonpool surface.

Aqua Lung – The first regulator invented by Jacques Cousteau and Emile Gagnan in 1943 which allowed divers to safely breathe from a bottle of compressed air worn on their back.

Aquanaut – Someone who lives and works undersea in a fixed habitat structure without resurfacing for a period of at least 24 consecutive hours.

Aquarius – The undersea habitat owned and operated by the US Government's National Oceanic and Atmospheric Agency (NOAA). It began operations in the US Virgin Islands in 1988 and is currently operating in the Florida Keys.

Aquatica – The combined region of the earth that is underwater and available for habitation by humanity. Aquatica includes both freshwater lakes and all of the planetary oceans.

Aquatican – A citizen of Aquatica, defined as an individual whose permanent residence is in an undersea or underwater habitat, base, outpost, colony or city.

Atlantica Expeditions – The activities of a group of individual explorers, initiated by the League of the New Worlds, whose purpose it is to establish the first permanent undersea colony.

Atmosphere (ATM) – A measure of pressure. One Atmosphere (1 ATM) is 14.7 pounds per square inch, 100 kilopascals ($\approx$750.062 torr) or 29.9230 inches of mercury.

Attenuation – The term for the process of exponential decrease in light's intensity with increasing depth.

Baylab – Morgan Wells' habitat operating in Chesapeake Bay that features a low energy life support system, one-of-a-kind end hatch, unique ultrasonic surface communications, and the capability of operations as the only self contained habitat ever built with no connection to the surface, if desired.

Bends – See Decompression Sickness.

Biome – The world's major ecological communities classified according to the predominant vegetation and characterized by adaptations of organisms to that particular environment. The largest of these is the earth's ocean or marine biome.

Bioregenerative Life Support – An advanced life support system that utilizes bioengineered systems to recycle nutrients, gasses and water, and in the process absorb wastes and produce food, oxygen and fresh water for the human consumer.

Biosphere – A biospheric system is one that is naturally balanced requiring little or no human intervention. Within the biosphere are balanced and synergistic components that both recycle and differentially sequester each part of gas, liquids and solids of the living and non-living constituent of the balanced ecology.

Biosphere II – A $200 million series of experiments conducted near Oracle, Arizona, in the late 20th century in a 3.15 acre enclosed biosphere that included human participants.

BioSUB – Australian Lloyd Godson's 2007 unique habitat creation that incorporated for the first time a bioregenerative component as a significant part of an undersea habitat's life support system.

Blue Water Habitat – An undersea habitat that is totally isolated from the surface except perhaps by cables carrying power, breathing gasses, water and communications (umbilicus). It is only accessible by direct diving or by submarine, diving bell or a diving bell that mimics an undersea elevator. It creates its own pressure and atmosphere and is specifically designed to indefinitely meet all aspects of life and living beneath the surface of the sea.

Carbon Dioxide Scrubber – A device into which air is circulated and carbon dioxide is filtered out by chemical or other physio-chemical / bioregenerative means.

CELSS Life Support – see Controlled Ecological Life Support System.

Challenger Station – A habitat being designed by LNW as the base and hub for a permanent human colony in the Gulf Stream set for deployment in 2012. Challenger Station will be the first truly hybrid, permanent, blue water habitat.

Claustrophobia – A fear or a phobia of confinement. A condition that causes the human to reflexively seek freedom from confinement. In an undersea habitat it is naturally triggered by elevated CO_2 levels, even with participants who have had no known prior episodes.

Color Absorption – In this discussion, it is the capacity of water to differentially absorb spectral energies (colors) with increasing depth. These colors are absorbed as depth increases in order of the first ones lost: red, orange, yellow, green, blue, indigo, violet. Note that these colors also represent the range of lowest spectral energies to highest.

Command Center – The location in the undersea habitat, base, outpost or colony in which all undersea operations are controlled and monitored.

Conshelf (Continental Shelf Station) – A series of experiments in undersea habitation by saturation diving techniques led by Jacques Yves Cousteau in the 1960's. It was described by Cousteau as a series of precursor experiments that would permit mankind a permanent presence on the seafloor of the world's continental shelves.

Continental Shelf – The vast areas of moderately deep waters that surround the world's continents. The continental shelves of the continental United States and Alaska consist of nearly 1.5 million square miles of seafloor, just less than half the territory of the United States.

Controlled Ecological Life Support System (CELSS) – Before 1970, early studies revolved around a 'Closed Ecological Life Support System' which developed over time and through research into the more appropriate definition of Controlled Ecological Life Support System. It is a system that markedly differs from a Biospheric System in that it is rigidly and extensively controlled and balanced by human intervention in all aspects. See also Bioregenerative Life Support.

Coriolis Force - The Coriolis force is a 'fictitious' force exerted on a body when it moves in a rotating reference frame. It is called a fictitious force because it is a by-product of measuring coordinates with respect to a rotating coordinate system as opposed to an actual 'push or pull.'

Current – Any more or less continuous, directed movement of ocean water that flows in one of the Earth's oceans. Currents are generated from the forces acting upon the water like the earth's rotation, wind, temperature and salinity differences and the gravitation of the moon. Depth contours, the shoreline and other currents influence the current's direction and strength.

Dalton's Law (also called Dalton's law of partial pressures) – states that the total pressure exerted by a gaseous mixture is equal to the sum of the partial pressures of each individual component in a gas mixture.

DCS – See Decompression Sickness.

Decompression – The process of gradual, safe relief of pressure on the human organism so that the aquanaut may return to the surface safely and uninjured without the effects of decompression sickness.

Decompression Sickness – The effect of a rapid release of pressure on an individual that results in bubbles of gasses (principally nitrogen) evolving in the blood stream. These bubbles can cause a myriad of ill effects from a mild rash to permanent paralysis to death by embolism.

Dry Suit – A diving suit, typically made of neoprene rubber, that is water tight and prevents cold water from coming into contact with the diver. It is sealed at the wrists and neck so that water does not enter the suit.

ELF (Extremely low frequency) – A broadcast technique, typically used by the military to communicate with its submarines while they are underwater. It is a very expensive form of communication that involves huge underground antenna arrays that are used to vibrate the crust of the earth. An ELF signal is typically broadcast at from 3 to 30 Hertz.

Excursion – When an aquanaut leaves the habitat and ventures outside under SCUBA or hookah.

Fathom – A nautical unit of measure of depth or length equal to six feet.

Habitat – A fixed undersea structure that provides both working accommodations and full life support capability to allow for extended or permanent human occupation by a single human or team of humans for the function of carrying out the various full ranges of processes that allow the human to live undersea. The habitat provides the total spectrum of living accommodations for all human functions from eating, sleeping and waste accommodations to socialization requirements.

Hardsuit – Also known as an ADS (Atmospheric Diving Suit), a hard shelled suit that isolates an individual from ocean water and pressure. It typically features an interior pressure of one atmosphere. It may also feature rotating limbs and end effectors to allow the diver to work isolated from the external pressure. Examples with movable legs are the JIM Suit (1969) and the Newt Suit (1987).

Hookah – A long hose with a typical SCUBA second stage regulator on one end that is attached to a large tank or air source connected to the habitat or surface based air source. It is a system that does not require the aquanaut to wear a tank or a personal air source for primary air.

Hybrid Habitat System – A hybrid habitat design may feature an ambient pressure section (such as the moonpool area and perhaps a submarine docking area) and a one-normal section for day to day activities.

Hybrid Life Support System – A hybrid life support system is one that may utilize two or several type components – simple and advanced or chemical and bioregenerative – to make up the total life support system design.

Hyperbaric – An environment whose pressure is greater than one atmosphere.

Hypercapnia – A physiological illness caused by elevated carbon dioxide levels. In its mild form it will produce symptoms of claustrophobia and in its more extreme forms it may produce nausea, dizziness, loss of consciousness and death.

Hypothermia – A physiological condition caused by a dramatic lowering of the body's temperature beyond which the body can control. It occurs in three stages and is manifested by shivering at stage one (body temperature may drop as low as 95 F); violent shivering and confusion at stage 2 (body temperature drops as low as 91.4 F); and at stage 3, cell function and major organs begin to fail, leading to death. Hypothermia occurs much more rapidly in water than in the air.

In Situ – Found in place or on site.

JIM Suit – 1 ATM ADS developed in 1969 by Underwater Marine Engineering. See Hardsuit.

Jules' Undersea Lodge – Originally the La Chalupa scientific research habitat. Refurbished into the world's first undersea hotel and opened for continuous business since 1986 in Key Largo, Florida.

Knot – 1.000 nautical mile per hour = 1.852 kilometer per hour. (A knot is never expressed as "knot per hour" since its units are implied.)

La Chalupa – A 3,300 cubic foot undersea scientific research habitat that operated off the coast of Puerto Rico in the 1970's. See Jules' Undersea Lodge.

League – Three (3) nautical miles.

League of the New Worlds (LNW) – A non-profit IRS 501(c)3 Scientific Research Corporation of Florida whose goal is commitment to the permanent human settlement of the ocean and space frontiers.

LED (Light Emitting Diode) Lighting – The LED light is an extremely versatile lighting system that allows for precise selection of spectral quality with a relatively low power requirement and low waste heat generation.

Leviathan – LNW habitat set for deployment in 2009 on the longest uninterrupted undersea mission and to test long duration undersea mission science, marine science and engineering concepts.

Life Support System (LSS) – A natural or human engineered system consisting of a group of devices that allow a human to survive in an environment hostile to human life.

LNW – See League of the New Worlds

Logistics – The science of planning, organizing and managing activities that provide goods or services.

Long Period Swell – Long-period swells are sometimes referred to as ground swells and have a period of 12-14 seconds or more. The longer the period of a swell, the deeper its effect will be felt.

L-POD (Logistic Pod) – A logistical transport case designed for the quick and efficient delivery of many kinds of goods to and from an undersea habitat.

LSS – See Life Support System

MarineLab – The undersea laboratory that currently holds the record for continuous operation of an undersea habitat. Originally designed by Neil Monney at the US Naval Academy and called *Medusa*. *MarineLab* began operating in 1985 in the Florida Keys and has hosted hundreds of individual investigations from pharmaceutical to space analog to military operations.

Mission Commander – The individual who is ultimately responsible for the safe and effective conduct of an undersea operation.

Mission Control – See Command Center.

Mixed Gas System – A life support system that mixes breathing gasses at percentages and components that are different than one-normal air. See NITROX.

Moonpool – The open hatch at the bottom of an ambient pressure habitat or an ambient pressure section of a hybrid habitat. The water is held out of the section by the pressure of the air inside which equals the pressure of the water at the opening.

NASA – National Aeronautics and Space Administration (a US Government Agency).

Nautical Mile – 1/60th of 1 degree of arc along the Earth's equator or a line of longitude. It works out to about 6,076 feet 01.386 inches (or exactly 1.852 km).

Newt Suit – 1 ATM ADS developed in 1987 by Dr. Phil Nuytten. See Hardsuit.

NextGen Habitat – A next generation habitat that is designed specifically for permanent human occupation.

Nitrogen Narcosis – A condition similar to intoxication with alcohol with euphoria, loss of balance and manual dexterity, disorientation and impaired reasoning that occurs in divers below 100 feet (30 meters) who breath compressed air because of the high nitrogen content of air. Nitrogen narcosis is reversed as the gas pressure decreases and the diver returns toward the surface.

NITROX – Any mixture of nitrogen and oxygen that contains less than the approximately 79% nitrogen as found in ordinary air. The most common use of NITROX mixtures containing higher than normal levels of oxygen is in SCUBA diving where the reduced percentage of nitrogen is advantageous in reducing nitrogen take up in the body's tissues and so extends the possible dive time and/or reduces the risk of decompression sickness.

NOAA – National Oceanic and Atmospheric Administration (a US Government Agency).

NURC – National Undersea Research Center at the University of North Carolina, Wilmington.

One-Normal Environment – An environment that consists of air made up of unaltered components of atmospheric air at one atmosphere pressure and one gravity at room temperature.

OSHA – Occupational Safety and Health Administration (a US Government Agency).

Oxygen Toxicity – The harmful effects caused by breathing oxygen at elevated partial pressures.

Partial Pressure – The pressure exerted by a single gas component in a mixture of gasses described by Dalton's Law. See also Dalton's Law.

Principal Investigator (PI) – In a scientific investigation, the Principal Investigator is the individual responsible for the conduct of the investigation.

Rebreather – A type of breathing set that provides a breathing gas containing oxygen and recycles exhaled gas.

Redundancy – In an engineering system redundancy describes the complete set of back-up systems that can assume any function of like design in the event of a system failure.

Regulator – A component of a SCUBA system that delivers breathing air to the aquanaut at a pressure that changes to equal the ambient pressure at the depth of the diver.

Resource Recovery – In older generation systems this was typically referred to as 'waste processing'. The resource recovery systems in advanced systems recycle and return the components of unwanted elements that may be produced in the system that, after processing, become the raw materials utilized in the next phase of the recycling process.

S3 (Surface Support System) – A permanent or semi-permanent surface presence with the purpose of providing services that can initially only be obtained at the surface and not undersea. A brief list of such services includes: power from solar, wind or ocean swell; an air source; communications; docking with surface vessels for personnel transfer; a platform for receiving helicopter aircraft; and all surface logistics connections to receive supplies.

Sacrificial Zinc Anode – A slug of elemental zinc that is bolted to the bare metal of a habitat and is attacked by the oxidative process preferentially, preserving the steel of the habitat so that it does not oxidize (rust) even while it is covered by seawater.

Saturation Diving – Diving in pressurized conditions for a period of time that will allow the body's fluids, including blood, to absorb dissolved gasses at the same partial pressures of the constituent gasses. Once the body is saturated, no more gas can enter the fluids and the diver is considered 'saturated' at that depth. At most depths, saturation is considered complete at approximately the nine hour point if the aquanaut's depth does not change.

Secchi Disk – An eight inch diameter disk divided into four quadrants painted black and white on opposite quadrants and used to determine undersea visibility.

Scott Carpenter Space Analog Station – A NASA undersea habitat designed and deployed at Key Largo, Florida, in 1997-1998.

Scouring – An erosion process resulting from the action of the flow of water.

SCUBA (Self Contained Underwater Breathing Apparatus) – Originally designed by Jacques Cousteau and Emile Gagnan in 1943. It is a suite of equipment that allows humans to breathe from its regulator underwater.

SEALAB – The United States Navy's series of undersea experiments in saturation diving during the 1960's.

Short Period Swell – Short-period swells are those with periods of 11 seconds or less.

Simple Life Support System – A life support system that does not recycle any part of its nutrients / resources back into the system.

Skins – A thin material diving suit that provides some thermal protection in cool water.

Statute Mile – 5,280 feet.

Submarine – A totally autonomous mobile undersea platform designed for the transport of humans.

Submariner – One who roams the seas in an underwater moving platform.

Submersible – A type of underwater vessel with limited mobility which is typically transported to its area of operation by a surface vessel or large submarine. Apart from size, the technical difference between a 'submersible' and a 'submarine' is that submersibles are not totally autonomous. They may rely on a support facility or vessel for charging of batteries, high pressure air, high pressure oxygen replenishment, or all of these.

Surface Connectivity – The type and degree to which an undersea habitat / colony is dependent on surface connections for any aspect of undersea life.

System Reliability – The probability that a system, including all hardware, firmware, and software, will satisfactorily perform the task for which it was designed or intended, for a specified time and in a specified environment

Tektite – An undersea habitation experiment featuring saturation in the late 1960's and early 1970's sponsored by General Electric, NASA and the Department of the Interior.

Umbilicus – The collection of insulated wires, cables, pipes and hoses that connect an undersea habitat to the S3 (Surface Support System).

Undersea Base – An undersea base is generally manned by explorers in a new area and functions as a place of operations as they survey the surrounding territory.

Undersea City – The undersea city differs from a colony in its range of services and the robustness of its infrastructure. An undersea city consists of residential, industrial and business areas together with administrative functions.

Undersea Colony – An undersea colony is one whose principal purpose is to establish its overall functions and purpose broad enough to include full family units. This uniqueness is what makes it different than a base or an outpost.

Undersea Habitat – A fixed undersea structure that provides both working accommodations and full life support capability to allow for extended or permanent human occupation by a single human or team of humans for the function of carrying out the various full ranges of processes that allow the human to live undersea. The habitat provides the total spectrum of living accommodations for all human functions from eating, sleeping and waste accommodations to socialization requirements.

Undersea Outpost – An undersea outpost is different than a 'base' in that it is an established presence that is returned to over and over and not typically abandoned and then totally reestablished at the end of each use. An outpost may (or may not) be 'manned' part time, but its infrastructure remains behind for the next users.

VHF (Very high Frequency) – This frequency band from 30 MHz to 300 MHz allows for line of sight communications and relative high data rate transfer.

Visibility – The limit to vision underwater. Visibility underwater may range from hundreds of feet to zero. Also referred to as viz.

Viz – See Visibility

Water Pressure – The pressure of the column of water as measured from the aquanaut to the surface. For every 33 feet of seawater depth the water pressure increases linearly at 14.7 pounds per square inch.

Wetroom – The room in which a diver enters a habitat. It is called a 'wetroom' because of the open moonpool and the room is characteristically damp from the frequent entry and exit of aquanauts. It is typically partitioned from the dry areas of the habitat to prevent excess humidity in the living areas. Wetroom areas may also include the habitat shower area.

Wetsuit - A neoprene or rubber garment of various thicknesses that traps water next to the diver's body, allowing the body's temperature to warm up the layer adjacent to the skin and provide some level of insulation against the cold, outside water. A wetsuit is effective to water temperatures down to around 45 degrees F, and below that a dry suit or hard suit is required for thermal protection.

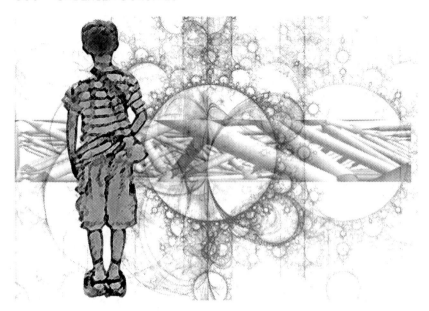

ABOUT THE AUTHOR

ennis Chamberland is an aquanaut explorer and undersea habitat designer who has logged more than 30 days living and working in undersea habitats.

As a young Native American boy growing up on the plains of Oklahoma, Dennis began a life's quest to permanently settle the oceans by building insect colonies in glass jars beneath a country stream that flowed near this home. Later, in 1972 at Oklahoma State University, he expanded the adventure in a campus wide experiment in undersea settlement titled the *Omega Project*. He was elected a Fellow of the New York Explorer's Club in 1991.

Dennis is a double alumnus of Oklahoma State University, earning his Bachelors Degree in 1974 and his Masters of Science Degree in Bioenvironmental Engineering in 1983. Dennis is a former United States Naval Officer and civilian Nuclear Engineer. He has professionally participated in the design of Advanced Life Support Systems considered for Moon and Mars bases.

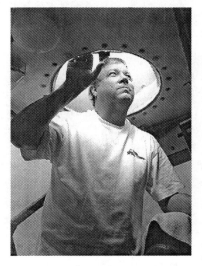

Dennis Chamberland founded the League of the New Worlds in 1991 as a not-for-profit IRS 501(c)3 Scientific Research Corporation of Florida whose goal is commitment to the permanent human settlement of the ocean and space frontiers. He is the Expedition's Leader for the Atlantica Expeditions whose purpose it is to establish the *Challenger Station* as the first permanent undersea colony off central Florida in 2012. The first phase of the Expeditions is scheduled to establish a new world's record of 80 days of continuous underwater habitation during 2009 in the *Leviathan* habitat.

Picking up where Dan Scott Taylor left off, Dennis is the Phase II Design Engineer for the 48 foot research submarine, the DST II, and will act as its Captain when it is launched.

Dennis Chamberland is also a passionate writer and speaker with 8 books and over 100 published works, including more than 30 entries in Magill's Survey of Science covering topics from ocean floor exploration to quantum physics and biological life support systems. Dennis is also a devoted family man and the father of six children. Working closely with his partner in life, Claudia Schealer Chamberland, Dennis has his sights pointed toward a permanent residence in Aquatica.

Please feel free to contact the author at

DennisChamberland.com

And review or purchase his other books at

QuantumEditions.com

ANNOTATIONS

History of Aquanauts Living and Working in the Sea
a. Bond, G., 1964, New Development in High Pressure Living,
Archives of Environmental Health, 9:310-314
b. Miller, J.W and Koblick, I.G., 1995, Living and Working In the Sea,
2nd ed.,P17
c. Bond, G., 1964, New Development in high Pressure Living,
Archives of Environmental Health, 9:311
d. Ibid., 314
e. Earle, S. and Giddings, A., 1980, Exploring the Deep Frontier,
National Geographic Society, P130
f. Ibid., P 129-130
g. Bennett, P.B. and Elliott, D.H., 1975, The Physiology and Medicine
of Diving and Compressed Work, 2nd Ed. Bailliere, Tindall and Cassell
h. Miller, J.W and Koblick, I.G., 1995, Living and Working In the Sea,
2nd Ed., P 61-62
i. NOAA Data, Aquarius Fact Sheet

The Great Abandonment
a. Great Lives from History, 1991, Salem Press, 487
b. Madsen, A., 1986, Cousteau – An Unauthorized Biography, P137

A Misdirected Start in Undersea Settlement
a. Marlwe, C., 2007, Career Self-Management in the New World of
Work, Internet Resource, P1
b. U.K. Sport Diving Commission Bulletin, May 2000, Internet
Resource, P1
c. Aitken, et.al.,International Journal of Andrology, April 23(2) P 116-
120

Going To Stay
a. Prospectus – League of the New Worlds, Inc., 1991, P1-10

The Undersea Environment
a. Helium Online Blog, "Project Atlantica – A Great Social
Experiment", 2007
b. Kolb, J.A., The Deep Sea, 2001, Institute of Marine Sciences, P115
c. Collins, S., 2007, How Do We Get Surf, Surfline.com, Internet
Resource

d. Wang, D.W., et. al., 2005, Extreme Waves Under Hurricane Ivan, Journal Science

Life Support Systems and Alien Atmospheres
a. Olstad, C.S., 1992, Seawater Exchange Systems To Provide Life Support In Manned Undersea Habitats, Marine Resources Development Foundation
b. Nuckols, M.L., 1999, Test and Evaluation of A Carbon Dioxide Absorption System Using Seawater For Underwater Life Support, US-Japan Cooperative Program in Natural Resources Panel on Diving Physiology, Tokyo
c. Chamberland, Dennis and Carpenter, Scott, 1996, Ocean Habitats as Analogs for Space Habitats, 26th International Conference on Environmental Systems, Monterey, California

Energy
a. Investigating the Ocean Gulf Stream, NOAA, Internet Resource
b. Alexander's Marvelous Machine, Spring 2005, National Resources Defense Council, One Earth Online Internet Resource

Darkness and Light
a. Franc D'Ambrosio of D'Ambrosio Architect , May 19, 2007, Canada.com

Economics
a. Chamberland, D. and Carpenter, S., 1996, Ocean Habitats as Analogs for Space Habitats, 26th International Conference on Environmental Systems, Monterey, 961397, P1-4

Politics
a. Ecology, 1992, 73(2), P713
b. Montaigne, F., Saving the Sea's Bounty, April 2007, National Geographic Magazine, PP36-97
c. Moore, C., November 2003, Natural History, 112(9)

Undersea City
a. Wikipedia Definition, Internet Resource

Undersea Empire
a. Ibid, Law of the Sea Treaty

WEB LINKS AND FURTHER EXPLORATION

Undersea Colonization

The Atlantica Expeditions – http://UnderseaColony.com

Yahoo Discussion Group –
http://tech.groups.yahoo.com/group/underseacolonies

Undersea Habitats in Current Use or in Operational Status

MarineLab Habitat – Deployed and operating continuously since in Key Largo, Florida since 1984. Its purpose is as an undersea laboratory primarily dedicated to marine science. Visit at: **http://www.mrdf.org/mul.htm**

Jules' Undersea Lodge - Deployed and operating continuously in Key Largo, Florida since 1986. Visit at: **http://www.jul.com/**

Baylab – Deployed and operational in Chesapeake Bay since 2000. Its primary mission is development of very low energy undersea life support systems and marine science research. *Baylab* is the only completely self-contained undersea habitat in the world. It has no power or gas lines connecting it to the surface. Visit at: **http://www.baylab.net/**

Aquarius - Deployed in the US Virgin Islands and began operations in 1988. After Hurricane Hugo in 1989, *Aquarius* was relocated to Wilmington, North Carolina where it was refurbished. Redeployed to the Florida Keys and resumed operations in 1993. Retreived for refurbishing in 1996 and resumed operations at its present location in 1998. Visit at: **http://www.uncw.edu/aquarius/**

OR

http://oceanexplorer.noaa.gov/technology/diving/aquarius/aquari
us.html

(Please be aware of some obvious and unfortunate **errors** made by the
web page designers in the very first words on the OceanExplorer
website.   They declare that Aquarius is "...the only undersea
laboratory dedicated to marine science operating in the world." And
atop the Google search engine for UNCW.edu, it is advertised as "The
world's only underwater habitat." Obviously, someone clearly forgot
to do their research. Hopefully the web engineers will correct these
embarrassing issues soon!)

Undersea Habitats in Construction

Poseidon Undersea Resorts – Set for deployment in Fiji. Check
site for opening date – it is shifting. Visit their remarkable site at:
http://www.poseidonresorts.com/

NOTE: Other undersea one atmosphere hotel projects are being
rumored from China to the Bahamas to Dubai. But no plans appear
firm and they seem to shift about – probably waiting on funding
sources.

The New Worlds Explorer – The New Worlds Explorer is a two
person – two day or "weekend" capable habitat that can be
launched from an ordinary boat trailer from most boat ramps. It is
scheduled for initial deployment in 2008 for use as an engineering
test habitat evaluating novel life support and undersea structural
component systems and evaluation as a recreational habitat. Visit
at: **http://UnderseaColony.com/nwe.html**

The Leviathan – Set for deployment in 2009 on the longest
uninterrupted undersea mission and to test long duration undersea
mission   science   and   engineering   concepts.   Visit   at:
http://UnderseaColony.com/leviathan.html

Undersea Habitats in Planning and Design Phase

The Challenger Station – Being designed as the first truly hybrid permanent blue water habitat. *Challenger Station* will be the base and hub for a permanent human colony off the coast of Central Florida in the Gulf Stream. Set for deployment in 2012. Visit at: **http://UnderseaColony.com/challsta.html**

British Columbia Institute of Technology Underwater Observatory – This habitat has just entered its first phases of design. If approved for construction, it will be used for BCIT students as an underwater laboratory in engineering and marine science.

BioSUB 2 – Lloyd Godson is reprising his fantastically successful and interesting experiment in Australia and is building a follow-up model to his original BioSUB. Current plans are to place the new habitat on the Great Barrier Reef. Lloyd is building the new habitat as part of a new television series to be produced and broadcast by a Canadian television network. Check out the BioSUB site at: **http://www.biosub.com.au**

Other Related Websites of Interest

Jaques Rougerie Architecte: **www.rougerie.com**
Sureda eSTUDIO, Guillermo Sureda-Burgos: **sureda.org**
Living and Working in the Sea – Miller and Koblick: **Amazon.com**
Dennis Chamberland books and ebooks: **QuantumEditions.com**
Personal Submarines News & Discussions: **www.Psubs.org**
Cambrian Foundation: **CambrianFoundation.org**
Marine Resources Development Foundation: **www.mrdf.org**
Undersea Colonies Discussion Group:
    http://tech.groups.yahoo.com/group/underseacolonies
League of the New Worlds: **ExpeditionsUSA.com**
Dennis Chamberland: **www.DennisChamberland.com**

THE GREAT ADVENTURE IS YOUR'S FOR THE ASKING!

Establishing the first permanent human colonies in Aquatica is our business. We are definitely looking for more than a few good people. We are looking for:

Engineers and Scientists of all types
Teachers of all levels, including elementary grades
Students of **all** grade levels and interests
Legal experts
Politicians
Environmentalists
Health Care Professionals
SCUBA Divers of all stripes and levels
And most other professions.

And of course, we are also looking to sign up people with a real passion to become

Permanent Colonists and Families

How do you become involved? Easy - you may apply here:

UnderseaColony.com/application - or -ExpeditionsUSA.com/application

You have the opportunity to become a Member-Explorer with the League of the New Worlds. We are dead serious about this. The first humans who leave for the colonies will come through our doors! There are many possible levels of involvement - they are all covered on the website. We look forward to having you on our team!

Dennis Chamberland
Expeditions Leader

Index

12 mile limit, 298, 331

Aaron Seven, 198, 336-345

accretion, 249-250

adapt, 5, 103, 104, 105, 108, 109, 110, 162, 164, 191, 192, 206

adaptability, 103, 109

advanced life support system, 125, 170, 265, 266, 355

air conditioner, 29, 30, 36, 168, 180, 181, 182

Alexander the Great, 54, 209

algae system, 94, 169, 170

Amazon River, 239

ambient habitat, 142, 150

Ambient Pressurized, 23

American Bureau of Shipping, 300

American Society of Mechanical Engineers, 300

Andrew, Hurricane, 145

Apollo, 167, 266, 310, 311

Apostle Paul, 53

Aqualung, 13, 59

Aquarius, 14, 87, 89, 91, 92, 93, 146, 172, 229, 356, 368, 370, 371

Aquatica, 3, 7, 17, 18, 19, 24, 25, 61, 89, 98, 107, 108, 112, 113, 114, 115, 124, 125, 130, 131, 136, 198, 220, 243, 249, 252, 254, 260, 269, 282, 304, 305, 308, 315, 321, 323, 332, 333, 339, 354, 356, 367

Aquatican(s), 7, 8, 18, 25, 90, 113, 114, 124, 125, 128, 129, 130, 131, 149, 150, 151, 202, 204, 241, 246, 249, 275, 282, 288, 308, 315, 332, 356, 3, 18,

19, 21, 112, 117, 122, 125, 128, 129, 130, 131, 136, 151, 192, 203, 205, 260, 266, 269, 274, 288, 332, 335, 351

artificial reef, 22, 145, 229, 248

Atlantic Silverside, 33

Atlantica, 159, 202, 353, 355, 356, 367, 368, 370

Atlantica Undersea Colony, 202

Atlantis Seafloor Colony, 222

attenuation, 135

Augustus Siebe, 58

Australian Broadcasting Company, 129

automated buoyancy control, 226

baralyme, 174

barnacles, 41

Baylab, 87, 93, 94, 356, 370

Bearing Sea, 140

Beebe, William, 58, 333

bends, 204

Bible, the, 52

biological diversity, 119

bioregenerative, 92, 94, 167, 168, 170, 264, 357, 359

Biosphere II, 285, 286, 287, 288, 289, 290, 357

Biospheric, 170, 358

BioSUB, 14, 94, 169, 357, 372

BioSUB 2, 94

Bishop, Joseph M., 38, 197

Black Sea, 142

black smokers, 19

Blue Water Undersea Habitat, 23, 24

Bluestriped Grunt, 35

Bond, George, 61, 66, 68, 70, 75, 76, 77, 105, 108, 352
Bonnicci, Alex Michael, 353
Bryan, Rob, 301, 353
Burgos, Guillermo Sureda, 203, 210, 353
Buttigieg, Ralph, 353
Cannon, Barry, 84, 96
Cambrian Foundation, 159
carbon dioxide - CO2, 29, 30, 33, 38, 66, 81, 84, 130, 151, 157, 158, 161, 164, 165, 167, 168, 169, 170, 171, 172, 174, 181, 234, 235,247, 267, 295, 326, 349, 357, 359
Carpenter, Scott, 14, 78, 79, 80, 92, 182, 353, 363
Carson, Rachel, 253, 254, 255, 256, 262, 269
catastrophic flooding, 185, 278, 280
Caution and Warning, 216
Cayman Salvage Master, 145, 147
CELSS, 91, 170, 357, 358
Challenger Station, 139, 261, 357, 367, 372
Charles Anthony Deane, 57
Chernomor, 85, 142, 143, 229
citizenship, 125
Chamberland, Claudia, 14, 31, 36, 37, 39, 40, 41, 43, 48, 116, 285, 303, 304, 308, 318, 320, 334, 353, 367
claustrophobia, 149, 150, 165, 359
Cocoa Beach, 297
Collisions, 283
Columbus, Christopher, 15, 271
Colvin Physical Education Center, 155

condensation, 179, 180
Conshelf I and II, 48, 68, 70, 72, 74, 75, 80, 81, 82, 96, 97, 98, 203, 214, 222, 315, 357
continental shelves, 70, 221, 222, 358
Cooper, Gordon, 80
Coriolis Force, 137, 358
Corriden, Darren, 37
countermeasure, 103, 104, 105, 176
Cousteau, Jacques-Yves, 13, 48, 59, 65, 68, 70, 72, 74, 75, 76, 80, 81, 95, 96, 97, 98, 99, 100, 101, 102, 105, 106, 108, 115, 153, 196, 203, 208, 222, 260, 355, 357, 363, 368
Dan Scott Taylor II - DST II, 172, 208,206, 311, 353
David, King, 53
Debua, Kurt, 8
decompression, 2, 23, 24, 33, 42, 62, 64, 66, 68, 75, 77, 81, 82, 91, 105, 108, 109, 110, 127, 142, 149, 157, 158, 159, 162, 202, 204, 277, 347, 358, 362
Deep House, 74
Deep Water Habitat, 24
dehumidification, 179, 180, 181, 182
Department of the Interior, 82, 364
Desalination, 242, 243
Diogenes, 70
Disney, 39, 317
Distillation, 242
Dive Saucer, 74
ecology, 40, 262, 357
Edison, Thomas, 121
Edmund Fitzgerald, 159

Edmund Halley, 56
Einstein, Albert, 128
electrolysis, 41, 250
emergency, 81, 149, 219, 225,
    235, 238, 240, 279, 280, 281,
    294
Emile Gagnan, 13, 59, 153, 355,
    363
English, Brett William, 304, 308
English, Eric Alexander, 304,
    308
EPCOT, 39, 258, 317
Exclusive Economic Zones,
    298, 330
exotic gas, 104, 163
Ezekiel, 53
failure, 10, 13, 120, 183, 185,
    220, 260, 272, 279, 287, 288,
    362
Fetch, 140
Fifth Principle
    of Undersea Exploration, 6
First Principle
    of Undersea Exploration, 5,
    108, 206, 306
Fourth Principle
    of Undersea Exploration, 6,
    253
free economy, 320
French Office of Underwater
    Research, 68
Friedman free market, 302
fuel cell, 240
Gemini, 80
General Electric Corporation,
    82, 317
Gentlemen Adventurers Club,
    205
Georges, Hurricane, 144
Gernhardt, Mike, 93

Godson, Lloyd, 14, 94, 129, 169,
    170, 313, 353, 357, 372
Graveline, Duane (Doc), 353
gravity, 18, 19, 104, 105, 221,
    222, 362
Guglielmo de Loreno, 56
Gulf Stream, 139, 239, 261, 262,
    357, 369, 372
gyres, 137, 139, 226
Hancock Seamount, 198, 224
hardsuits, 110, 111, 159
heat exchanger, 169, 179, 182,
    191
Heavy metals, 267
Helgoland, 85
helium, 75, 76, 77, 80, 81, 82,
    105, 111
Henry Fleuss, 58
Hilbertz, Wolf, 249
Hillary list, 115
Hillary, Sir Edmund, 114, 115,
    116, 118
HMS Eagle, 57
Holland, Al, 90
Hollywood, 315, 316
Homer, 54
hookah, 31, 35, 81, 359
Horizons EPCOT, 317, 318
Humboldt, 139
humidity, 29, 38, 74, 80, 157,
    161, 168, 175, 179, 180, 181,
    234, 235, 243, 308, 365
hurricane, 91, 144, 145, 146,
    147, 240, 282, 288
Hybrid Habitat, 22, 150, 359,
    361
hydrodynamic force, 146
Hydrolab, 82, 83, 87, 89, 122,
    229, 311
hydrophilic tubes, 235

Hydrosleds, 211
hydrostatic pressure, 231
hypercapnia, 164, 165
Iliad, 54
immunosuppression, 104
in situ, 6, 118, 125, 172, 215, 240, 244, 246, 247, 252
insulation, 179, 181, 365
Irene, Hurricane, 144
Ivan, Hurricane, 146, 369
Jim Atria, 145, 147
John Lethbridge, 56
Jones, Bruce, 196
Jules' Undersea Lodge, 8, 14, 44, 48, 87, 89, 122, 279, 311, 320, 360, 370
Katrina, Hurricane, 145
Kennedy, John F., 60, 148
Key Largo, 8, 14, 15, 36, 89, 91, 92, 146, 305, 311, 320, 360, 363, 370
Key West, 75, 145
King Ferdinand, 271
King John II, 271
Kings Point Merchant Marine Academy, 38
Koblick, Ian, 8, 44, 48, 67, 76, 95, 351
La Chalupa, 8, 85, 87, 89, 90, 230, 360
La Jolla, California, 77, 256
La Pression Barometrique, 58
latent stress, 149, 150
League of the New Worlds Corporation, 130
LED lighting, 191, 236
Lenenger, Jerry, 281
L'Enfant, Pierre, 314
Leonardo da Vinci, 56

Leviathan, 172, 314, 353, 360, 367, 371
light pipe, 236
Lindbergh, Jon, 75
Lindbergh, Charles, 75
Link, Edwin A., 66
lithium hydroxide, 174
Living Seas, 39, 203
LNW, 255, 311, 355, 357, 360
London Company, 274
long period swell, 141
Logistics Poid, L-POD, 218, 219, 361
Life Support System, LSS, 157, 158, 161, 166, 167, 168, 169, 170, 174, 234, 360, 361
Man In The Sea, 68
Mandeville, Bill, 47
Manta Ray anchors, 231
Marine Resources Development Foundation, 36, 89, 369
MarineLab, 8, 14, 42, 48, 87, 89, 90, 91, 92, 122, 123, 181, 279, 311, 361, 370
Mark V Dive Helmet, 58
McMurdo Station, 312
Medusa, 87, 361
Mercedes, 145, 147
Mercury astronauts, 270
metabolism, 29, 164, 171
microwave, 36, 38, 39, 46, 47, 168, 236
military, 53, 54, 57, 58, 61, 62, 66, 76, 78, 108, 120, 186, 191, 214, 219, 236, 279, 294, 312, 331, 332, 338, 359, 361
Miller, James W., 69, 76, 82, 230, 351
Miller, Steve, 144
mining, 120, 247, 257, 291, 322

Mir, 92, 281
Mission Commander, 36, 37, 47, 49, 150, 275, 276, 282, 361
Mission Control, 36, 37, 49, 361
mixed gas, 109, 158, 160, 317, 346
Monterey Bay Aquarium, 39
Moon, 133, 258, 366
moonpool, 23, 28, 30, 31, 32, 33, 35, 36, 44, 46, 48, 50, 94, 103, 105, 110, 142, 143, 148, 150, 165, 185, 187, 204, 208, 209, 211, 279, 316, 355, 359, 365
Mudd, Vicki, 354
NASCAR, 274
National Atmospheric and Atmospheric Administration - NOAA, 14, 82, 89, 91, 92, 93, 144, 146, 172, 356, 362, 368, 369
National Geographic Society, 68, 368
National Underwater Research Center, 144
NAUI, 60
Navy Seal, 159
NEDU, 93
NEEMO, 93
Next Generation - NEXT GEN, 166, 168, 197, 219, 361
nitrogen narcosis, 76, 163
North Atlantic, 7, 137, 178, 262, 336, 340
North Pacific Drift, 139
Northwest Passage, 103
Novak, Brian, 320, 353
Novak, Eric, 304, 308, 321, 353
Novak, Ian, 3
Novak, Shelley, 3, 320, 323, 353
Novak, Tim, 320, 353

Nuckols, M.L., 172
NURC, 92, 93, 144, 362
Ocean Engineering Test Range, 83, 84
OCEAN Project, 91
oceanography, 43, 274
Oklahoma, 8, 16, 102, 153, 155, 366
Oklahoma State University, 8, 102, 155, 366
Olstad Chris, 8, 90, 124, 171, 181
Omega Project, 8, 366
one atmosphere habitat, 172
O'Neill, Gerard K., 222
Occupational Safety and Health Administration - OSHA, 161, 362
outpost, 125, 266, 310, 312, 313, 314, 322, 324, 325, 355, 356, 357, 364, 365
oxygen, 30, 58, 62, 66, 75, 76, 81, 125, 130, 151, 156, 157, 158, 160, 163, 164, 165, 168, 169, 170, 171, 178, 202, 215, 235, 246, 247, 250, 262, 281, 287, 295, 326, 349, 356, 362, 364
Pacific, 10, 11, 77, 138, 139, 140, 198, 292, 293
Pacifica, 198, 199, 216, 224
PADI, 60
Pangea, 17
Permanence, 234, 282
photosynthesis, 169
Plankton, 262
Platform Mounted Habitat, 22
Platipus anchors, 231
Plinius the Elder, 56
Plymouth Company, 274
Pollution, 118, 293

Polyakov, Valery, 90

Port Sudan, 74

Presley, Richard, 90

Principal Investigator, 150, 362

Problemata, 54

professional divers, 159, 160, 205, 347

Project Atlantis, 90

Project Genesis, 64

Quantum Storms, 198, 216, 224

Queen Isabella, 271

Receiving and Docking, 215

redundancy, 167, 362

regulator, 29, 32, 34, 155, 156, 355, 359, 363

research, 43, 64, 98, 119, 122, 130, 138, 167, 211, 218, 249, 274, 326, 358, 360, 370, 371

Resort Habitat, 22

resource recovery, 161, 242, 243, 247, 263,264, 265, 266, 269, 362

Rougerie, Jacques, 196

Surface Support System - S3, 171, 178, 208, 209, 212, 213, 214, 215, 216, 217, 218, 238, 283, 336, 337, 339, 363, 364

sacrificial zink, 42

Safety Standard for Pressure Vessels for Human Occupancy, 300

Saltstraumen Sound, 139

Salyut space station, 181

San Clemente Island, California, 83

Scala, Chris, 250

Schealer, Richard B., 47

scientists, 10, 45, 49, 84, 89, 118, 119, 158, 255, 274, 275, 289, 305

Scott Carpenter Space Analog Station
SCSAS, 14, 92, 182, 363

scouring, 143, 144, 147

scrubber, 30, 172, 235, 326

SCUBA, 20, 21, 56, 59, 89, 91, 117, 133, 134, 151, 153, 154, 155, 156, 157, 159, 163, 204, 205, 208, 278, 304, 317, 324, 359, 362, 363

Scyllias, 54

Sea Base Alpha, 203

SEA TEST, 92

SEALAB I, 75, 76, 77, 78, 80

SEALAB II, 77, 78, 80, 84, 178, 238

SEALAB III, 84, 97

Seament, 249-250

seasick, 142

Seasonal Affective Disorders - SAD, 189

Seatopia, 85

seawater hydrolysis, 240

Second Principle of Undersea Exploration, 5, 209, 212, 239

Shallow Water Habitat, 24

sharks, 350

Silent Spring, 255

smart systems, 168

Smithsonian Institution, 68, 82

Society of Merchant Venturers, 274

sonar, 283, 284, 343

sovereignty, 246, 331, 334

Soviet, 96, 142, 181

space analog, 43, 92, 274, 361

Space Exploration, 103

spectral energy, 135

spectral quality, 190, 360

Submerged Portable Inflatable
  Dwelling - SPID, 75
Starfish House, 48, 72, 74, 76,
  203
Stenuit, Robert, 66, 68, 70, 75,
  78
stewards, 13, 118, 254
Stonebrooke, Tennessee, 4, 202,
  233
suction pile, 231
supercritical air, 171
Surface Support System, 178
  S3, 171, 212, 213, 363, 364
suspended colony, 224, 226
swell, 140, 141, 142, 143, 144,
  146, 209, 213, 215, 216, 224,
  225, 239, 247, 337, 342, 361,
  363
Swell Generator, 216
system reliability, 167
The Edge of the Sea, 255
The Sea Around Us, 254, 255,
  256
Third Principle
  of Undersea Exploration, 6,
  246
Third United Nations
  Conference on the Law of
  the Sea, 330
Thucydides, 54
Titanic, 41, 178
Todd, Bill, 92, 93
tort, 300, 301, 302
Trieste, 60, 186
Turtle, 57
typhoon, 282
Tysall, Terrence, 1, 159
undersea city, 314, 325, 326, 328,
  364
undersea colony, 6, 18, 74, 125,
  130, 139, 186, 189, 198, 201,
  202, 203, 204, 205, 206, 208,
  209, 211, 212, 214, 215, 216,
  217, 218, 226, 233, 234, 236,
  237, 238, 239, 240, 242, 243,
  244, 246, 248, 254, 258, 259,
  261, 266, 267, 271, 274, 276,
  277, 278, 280, 281, 282, 284,
  309, 313, 314, 315, 317, 318,
  319, 320, 321, 324, 328, 336,
  337, 345, 356, 364, 367
undersea habitat, 15, 21, 22, 24,
  29, 37, 39, 49, 87, 90, 94, 110,
  124, 130, 145, 146, 148, 156,
  157, 164, 167, 169, 178, 183,
  185, 189, 191, 194, 197, 202,
  204, 214, 216, 234, 279, 317,
  346, 349, 350, 351, 355, 356,
  357, 361, 363, 364, 366, 370
Undersea Scooters, 211
undersea station, 119, 294
Underwater Kinetics, 219
Underwater Vehicles, Systems
  and Hyperbaric Facilities,
  300
United States Constitution, 298
US Divers, 154
US Naval Proceedings, 79
USS *Nautilus*, 66
USS *Spiegel Grove*, 145, 147
Vaspucci, Amerigo, 25
Vind, H.P., 171
Virginia Company, 274
visibility, 34, 35, 74, 80, 130,
  135, 136, 137, 138, 139, 160,
  180, 184, 219, 363
viz, 137, 138, 160, 365
Von Guericke, 56
Waldseem, Martin, 25
Walt Disney, 203, 257
Walt Disney World, 203

Ward, Mark E., 90, 92, 93
Washington D.C., 313
wave crest, 141
wave period, 141, 147
wavelength, 135, 140, 141, 142
Wells, Morgan, 71, 93, 356

Wind power, 215
Windows, 180, 185, 186
YMCA, 59, 154, 155
zinc. *See* Sacrificial Zinc
zoology, 43

Printed in the United States
102354LV00002B/25/A

9 781889 422152